Cambridge Elements

Elements in Climate Change and Cities: Third Assessment Report of the Urban Climate Change Research Network

Series Editors

William Solecki, *New York*

Minal Pathak, *Ahmedabad*

Martha Barata, *Rio de Janeiro*

Aliyu Salisu Barau, *Kano*

Maria Dombrov, *New York*

Cynthia Rosenzweig, *New York*

PLANNING, URBAN DESIGN, AND ARCHITECTURE FOR CLIMATE ACTION

Coordinating Lead Authors

Jeffrey Raven, *New York*

Mattia Federico Leone, *Naples*

Lead Authors

Sanjukkta Bhaduri, *New Delhi*

Christian Braneon, *New York*

David Corbett, *Munich*

David Driskell, *Seattle*

Ursula Eicker, *Montréal*

John Fernández, *Cambridge*

Jing Gan, *Shanghai*

Anna Hürlimann, *Melbourne*

Ilana Judah, *New York*

Michael Neuman, *London*

Barbara Norman, *Canberra*

Dennis Pamlin, *Stockholm*

Chao Ren, *Hong Kong*

Rob Roggema, *Monterrey*

Pourya Salehi, *Bonn*

Anne Shellum, *New York*

Andréa Souza Santos, *Rio de Janeiro*

Joel Towers, *New York*

Cristina Visconti, *Naples*

Shaftesbury Road, Cambridge CB2 8EA, United Kingdom

One Liberty Plaza, 20th Floor, New York, NY 10006, USA

477 Williamstown Road, Port Melbourne, VIC 3207, Australia

314–321, 3rd Floor, Plot 3, Splendor Forum, Jasola District Centre,
New Delhi – 110025, India

103 Penang Road, #05–06/07, Visioncrest Commercial, Singapore 238467

Cambridge University Press is part of Cambridge University Press & Assessment,
a department of the University of Cambridge.

We share the University's mission to contribute to society through the pursuit of
education, learning and research at the highest international levels of excellence.

www.cambridge.org
Information on this title: www.cambridge.org/9781009643917

DOI: 10.1017/9781009643894

First published 2025

A catalogue record for this publication is available from the British Library

ISBN 978-1-009-64391-7 Hardback
ISBN 978-1-009-64393-1 Paperback
ISSN 2976-9116 (online)
ISSN 2976-9108 (print)

Additional resources for this publication at www.cambridge.org/raven-et-al

Planning, Urban Design, and Architecture for Climate Action

Elements in Climate Change and Cities: Third Assessment Report of the Urban Climate Change Research Network

DOI: 10.1017/9781009643894
First published online: November 2025

Coordinating Lead Authors
Jeffrey Raven, *New York*   Mattia Federico Leone, *Naples*

Lead Authors

Sanjukkta Bhaduri, *New Delhi*
Christian Braneon, *New York*
David Corbett, *Munich*
David Driskell, *Seattle*
Ursula Eicker, *Montréal*
John Fernández, *Cambridge*
Jing Gan, *Shanghai*
Anna Hürlimann, *Melbourne*
Ilana Judah, *New York*
Michael Neuman, *London*

Barbara Norman, *Canberra*
Dennis Pamlin, *Stockholm*
Chao Ren, *Hong Kong*
Rob Roggema, *Monterrey*
Pourya Salehi, *Bonn*
Anne Shellum, *New York*
Andréa Souza Santos, *Rio de Janeiro*
Joel Towers, *New York*
Cristina Visconti, *Naples*

Author for correspondence: Jeffrey Raven, jraven@nyit.edu

Abstract: Embedding climate-resilient development principles in planning, urban design, and architecture means ensuring that transformation of the built environment helps achieve carbon neutrality, effective adaptation, and well-being for people and nature. Planners, urban designers, and architects are called to bridge the domains of research and practice and evolve their agency and capacity, developing methods and tools consistent across spatial scales to ensure the convergence of climate actions. Shaping change necessitates an innovative action-driven framework with multidimensional analysis of urban climate factors and co-mapping, co-design, and co-evaluation with city stakeholders and communities. This Element provides an assessment on urban climate factors, system efficiency, form and layout, building envelope and surface materials, and green/blue infrastructure and how they affect key metrics and indicators related to complementary aspects including greenhouse gas emissions, impacts of extreme weather events, spatial and environmental justice, and human comfort. This title is also available as open access on Cambridge Core.

Keywords: built environment, urban planning, urban design, architecture, climate justice

ISBNs: 9781009643917 (HB), 9781009643931 (PB), 9781009643894 (OC)
ISSNs: 2976-9116 (online), 2976–9108 (print)

Contents

Additional resources for this publication, including
a Glossary, are available at www.cambridge.org/raven-et-al

Contributors[*]

Contributing/Case Study Authors

Sara Artaega-Morales
Apartadó
Ione Avila-Palencia
Philadelphia
Jole Lutzu
Barcelona
Bruno Barroca
Paris
Martina Kohler
New York
Margot Pellegrino
Paris
Juliana Velez Duque
Bogotá
Darien Alexander Williams
Cambridge
Sean O'Donoghue
Durban
Maria Iaccarino
Naples
Enza Tersigni
Naples
Aliyu Salisu Barau
Kano
Enjoli Dominique Hall
Cambridge
Daphne Gondhalekar
Munich

Element Shepherds

Ken Doust
Lismore

---

[*] ARC3 authors are associated solely with their cities or metropolitan regions.

Martin Lehmann
Aalborg
Franco Montalto
Philadelphia
Juan Camilo Osorio
New York

Element Scientists[†]

Enza Tersigni
Naples
Mia Prall
Aalborg
Priska Marianne
Jakarta
Colin Marchenoir
London

---

† ARC3 Element Scientists support the Coordinating Lead Authors in content development and research support.

Series Preface

Urban Climate Change Research Network

Third Assessment Report on Climate Change and Cities (ARC3.3)

William Solecki (New York), Minal Pathak (Ahmedabad), Martha Barata (Rio de Janeiro), Aliyu Salisu Barau (Kano), Maria Dombrov (New York), and Cynthia Rosenzweig (New York)

Cities and the urbanization process itself are at a crossroads. While the world's urban population continues to grow, cities are increasingly pressed by chronic and acute stresses including increasing inequity, polluted air and waters, limited governance and financial capacities, along with entrenched spasmodic crime and conflict, and the COVID-19 pandemic. Climate change has now exacerbated these problems and in many cases created new ones, at a time when cities are asked to be the bulwark of the climate-solution space. The advent and application of new technologies and strategies associated with the internet, environmental sensing, multi-modal transport, and innovative planning and design strategies portend a new golden age for cities. Some cities provide glimmers of this possible future, but persistent stresses and intermittent crises, along with climate change, push against progress. In the *Third Assessment Report on Climate Change and Cities* (ARC3.3) of Urban Climate Change Research Network (UCCRN), we directly address these issues head-on and present state-of-the-art knowledge on how to bring all cities and their residents forward to a more sustainable future.

An absolute necessity now exists for cities everywhere, both in the Global North and Global South, to aggressively work to fulfill their potential as leaders in climate change action. In the Global North, the task is for cities to address the emerging challenges from the changing climate and the exigencies of compliance with the UNFCCC Paris Agreement. For cities in the Global South, there is the double burden of achieving climate-resilient development, that is, meeting increasing demand for housing, energy, and infrastructure for burgeoning populations, while confronting simultaneous requirements to reduce greenhouse gas emissions and adapt to a changing climate (UNEP & UN-Habitat, 2021). In all geographies, the implementation of transformative mitigation and adaptation in cities can be an instrument to generate livelihoods for those with lower purchasing power and can enhance capacity to better respond to shocks such as future pandemics, energy supply chain spasms, and food security emergencies (UNDP, 2022).

Benchmarked Learning

ARC3.3 builds upon the preceding *UCCRN Assessment Reports on Climate Change and Cities*, ARC3.1 (2011), and ARC3.2 (2018). The purpose of the ARC3 series is to provide the benchmarked knowledge base for cities as they affirm their essential responsibility as climate change leaders. The ARC3 Series, with newly added ARC3.3 Elements, presents knowledge that builds on accumulated and shared experiences and thus advances and deepens with time.

In ARC3.1, cities were identified as key actors – "first responders" – in rising to the challenges posed by climate change (Rosenzweig et al., 2011). According to ARC3.1, "Cities around the world are highly vulnerable to climate change but have great potential to lead on both adaptation and mitigation efforts."

In ARC3.2, this focus advanced into understanding of how cities can achieve their potential by establishing a multifaceted pathway to transformation (Rosenzweig et al., 2018). It provided a roadmap for cities to fulfill their leadership potential in responding to climate change. According to ARC3.2, "As cities mitigate the causes of climate change and adapt to new climate conditions, profound changes will be required in urban energy, transportation, water use, land use, ecosystems, growth patterns, consumption, and lifestyles."

Now, as the urgency of climate change is brought home daily, ARC3.3 offers the knowledge needed to *speed up and scale up* urban action on climate change. To accomplish this, ARC3.3 presents practical methods and case study examples for accelerating change into rapid transformation in cities across the globe.

UCCRN Assessment Process

ARC3.3 authors were either self-nominated or nominated by a third party and were selected by the ARC3.3 Editorial Board through comprehensive vetting that prioritizes expertise, diversity, gender, and geographic balance. Each Author Team develops a robust assessment of an Element topic, using the latest literature, while also conducting new research. All Author Teams are responsible for conducting a stakeholder engagement session during the writing period, with the goal of ensuring relevance to a diverse group of urban decision-makers. During self-coined "Stakeholder Soundings," authors present emerging major findings and key messages to stakeholders, including city leaders from the authors' home cities, for their feedback. UCCRN also coordinates a rigorous iterative peer-review process for each ARC3.3 Element that engages with both academic and practitioner experts, both in and out of the network.

UCCRN's Case Study Docking Station (CSDS) is a searchable database designed to facilitate peer-learning between and among cities, benchmark actions over time, and enable cross comparisons.[1] The CSDS includes more than 200 peer-reviewed case studies covering a range of topics such as climate change vulnerability, hazards and impacts, and mitigation and adaptation actions for specific sectors. The CSDS has a total of sixteen searchable variables.[2] For example, users can filter searches by climate zone, city population size, human development index, gross national income, and mitigation versus adaptation, or directly type in keywords and city names. Case study examples include flood adaptation in Bridgetown, cloudburst planning in Copenhagen, and climate action financing in Durban.

Cities are vanguard sites for opportunities to enhance equity and inclusion. Besides ARC3.3 *Justice for Resilient Development in Climate-Stressed Cities*, equity and inclusion permeate every ARC3.3 Element, as city experts delve into the multiple dimensions of climate change justice: distributive (relating to differential vulnerability of groups and neighborhoods), contextual (relating to the root causes of vulnerability), and procedural (relating to participation in decision-making for climate change interventions) (Foster et al., 2019). Recognitional (valuing of diverse identities) and restorative justice (restoring dignity and repairing the societal harm caused by earlier actions) concepts are now emerging, as well. Elucidation of ways to achieve all types of climate justice for the most vulnerable urban groups and equal access to financial and technological resources for all cities underpins ARC3.3.

ARC3.3 Elements

UCCRN has conducted city-centered assessments since its founding in 2007. With more than 2,000 scholars and experts from cities around the world, UCCRN is addressing the research agenda that was formulated at the Intergovernmental Panel on Climate Change (IPCC) Conference on Cities and Climate Conference (Prieur-Richard et al., 2018).[3] Key components of

---

[1] UCCRN is developing the City Solutions Case Study Atlas (City CSA) which is an accessible database of climate change adaptation and mitigation strategies that cities globally are implementing within their local contexts. Case studies included in the UCCRN City CSA pay special attention to solutions-based projects and initiatives in urban and metropolitan settings.

[2] The searchable variables available to users on the UCCRN CSDS are: ARC3 Assessment, Language, Continent, Country, City, Urban Design Climate Workshop, Köppen Climate Zone, Coastal (marine or riverine), City Population Classification, Urban Density, GNI Classification, HDI Index, Gini Index Coefficient, Type of Climate Intervention, Keywords. https://uccrn.ei .columbia.edu/case-studies.

[3] The Urban Climate Change Research Network actively participates in conferences that highlight the role of cities in climate change such as Innovate4Cities, the World Urban Forum, and Adaptation Futures.

this research agenda include urban planning and urban design; green and blue infrastructure; equity, health, and sustainable production and consumption; and finance. More than 300 UCCRN authors have now advanced this research agenda and other critical topics through the *Third Assessment Report on Climate Change and Cities*, which consists of twelve peer-reviewed monographs published as Cambridge University Press Elements, both separately and together, throughout 2025 and 2026.[4]

1. Learning from COVID-19 for Climate-Ready Urban Transformation

The COVID-19 pandemic has revealed gaps in city readiness for simultaneous responses to pandemics and climate change, particularly in the Global South. However, these concurrent challenges present opportunities to reformulate current urbanization patterns, economies, and the dynamics they enable. This Element focuses on understanding COVID-19's impact on city systems related to mitigation and adaptation, and vice versa, in terms of warnings, lessons learned, and calls to action.

2. Justice for Climate-Resilient Development in Climate-Stressed Cities

To ensure climate-resilient urban development, both adaptation and mitigation must include the broader city contexts related to equity, informality, and justice. Responses to climatic events are conditioned by the informality of the existing social fabric, institutions, and activities, and by the inequitable distribution of impacts, decision-making, and outcomes. This Element elucidates differential exposure to climate events, as well as distributive, recognitional, procedural, and restorative justice.

3. Planning, Urban Design, and Architecture for Climate Action

Architects, urban designers, and planners are called on to bridge the domains of research and practice and evolve their capacity and agency by developing new methods and tools consistent across multiple spatial scales. These are required to ensure the convergence of effective outcomes across metropolitan regions, cities, neighborhoods, and buildings. This Element evaluates how the fields of urban planning, urban design, and architecture can integrate climate mitigation and adaptation and presents a manifesto for urban transformation using science-informed methods and tools.

4. Financing Urban Transitions to Climate Neutrality and Increased Resilience in Cities

This Element assesses the availability of, and access to, finance for mitigation and adaptation in urban areas. It evaluates current international flows, national

---

[4] See www.cambridge.org/core/publications/elements/elements-in-climate-change-and-cities for the full set of ARC3.3 Elements and authors.

policies, and municipal utilization capacities across private and public sectors, and nongovernmental and community-based organizations. Global financial capital is abundant but often flows to corporate investments and real estate development rather than into critical efforts to mitigate and adapt to climate change in cities. Political will and public pressure are crucial to effectively redirect these funds.

5. Urban Climate Science: Building the Knowledge Base for City Risk Assessments and Resilience

Cities alter the climate system both within their boundaries and nearby through interactions with impervious land surfaces, energy generation, and transportation systems. These processes that occur on urban scales are interacting with larger-scale climate change to exacerbate extreme events that impact urban dwellers. This Element provides observations and projections of temperature, precipitation, and sea-level change for the cities engaged in ARC3.3. It assesses the latest research on heat and precipitation islands, compound extreme events, and indicators and monitoring, including the use of remote sensing in urban settings.

6. Governance, Enabling Policy Environments, and Just Transitions

The nature of governance, as a concatenation of social institutions and practices embedded at different scales, suggests the need for an integrated approach to address the complex challenges of climate change in cities. This Element sets forth multi-level governance (MLG) structures for climate action across urban, provincial, national, and international levels, analyzes the inclusion of urban actions in Nationally Determined Contributions (NDCs), and assesses the potential for urban transitions and transformations.

7. Infrastructure for a Net Zero and Resilient Future in Cities

Without infrastructure, cities could not exist. Infrastructure determines urban form, functions, economic development, people's livelihoods, and well-being. By developing transformative infrastructure, cities can achieve ambitious greenhouse gas emission reductions, build resilience to climate impacts, and ensure inclusive and diverse access to services. This Element explores infrastructure planning concepts including life cycle analysis, decentralization, and integration, and emphasizes the need for equitable, resilient systems designed according to future climate projections.

8. Nature-Based Solutions: Enhancing Capacity to Respond to Shocks and Stresses

There is a growing acknowledgment that a disproportionate amount of attention and finance is invested in hard infrastructure to mitigate and adapt to climate change in cities. In contrast, soft infrastructure, that is, the use of natural features and processes, has been comparatively overlooked until recently. This Element assesses the ways that nature-based solutions (NbS) – such as reforestation, urban parks, street trees, sustainable urban drainage systems, and community gardens – can enhance the capacity of cities to reduce greenhouse gas emissions and enhance resilience to climate stresses.

9. Circular Economies for Cities

Circularity, an economic system where waste and pollution is minimized and resources are continuously reused, has the potential to transform cities and city systems in both the Global North and the Global South. Sustainable consumption and production and supply and demand factors are increasingly being analyzed in urban contexts. This Element addresses the linkages of circular economics to climate action planning, the water–energy–food system nexus, and just, local development.

10. Data and the Role of Technology

Over the past decade, changes in internet penetration and the development of new information and communication technologies have catalyzed an ecosystem of approaches that employ "big data" and "smart tools" to support adaptation and mitigation. Artificial intelligence and machine learning play a large role in this new technological ecosystem. This Element evaluates the opportunities and challenges for cities as they employ these new technologies and assesses emerging tools for their utility in climate change responses.

11. Perception, Communication, and Behavior

This Element explores the latest research on how urban residents perceive climate change so that effectiveness of actions can be improved. An important corollary to this is the role that communication plays in how mitigation and adaptation actions are adopted by cities. In the event of a climate disaster, the way that cities communicate has a direct effect on residents' perception of risks and subsequent behaviors such as evacuation or strategic relocation. The Element addresses how behaviors by urban inhabitants can be encouraged to change mobility patterns and energy use in order to reduce greenhouse gas emissions, while simultaneously helping citizens to prepare for increasing climate extremes.

12. Health and Well-Being

Climate change, especially increasing extreme events, are exacerbating the risks of mortality, disease, injury, and impacts on physical and mental health and well-being in many cities. Climate change also has indirect impacts on health through disruptions in food supply and water availability. This Element assesses the latest findings on all aspects of the intersection of health and climate change for urban residents – including built form and presence of natural spaces.

ARC3.3 Major Findings and Key Messages

Besides the basic assessment content, each Element includes a statement of Major Findings and Key Messages. Major Findings bring forward significant new knowledge that emerged through the assessment process, while Key Messages are recommendations for new directions, plans, and activities with a specific focus on opportunities to speed up and scale up urban climate action.

Cross-Cutting Themes

Cities are complex social-ecological-technological systems. While ARC3.3 is composed of twelve separate Elements, together they comprise multiple synergies, interdependencies, and points of intersection. To highlight these connections, each Element addresses its own selection of relevant cross-cutting themes (CCTs). Figure 1 illustrates how significant recurring themes appear within an

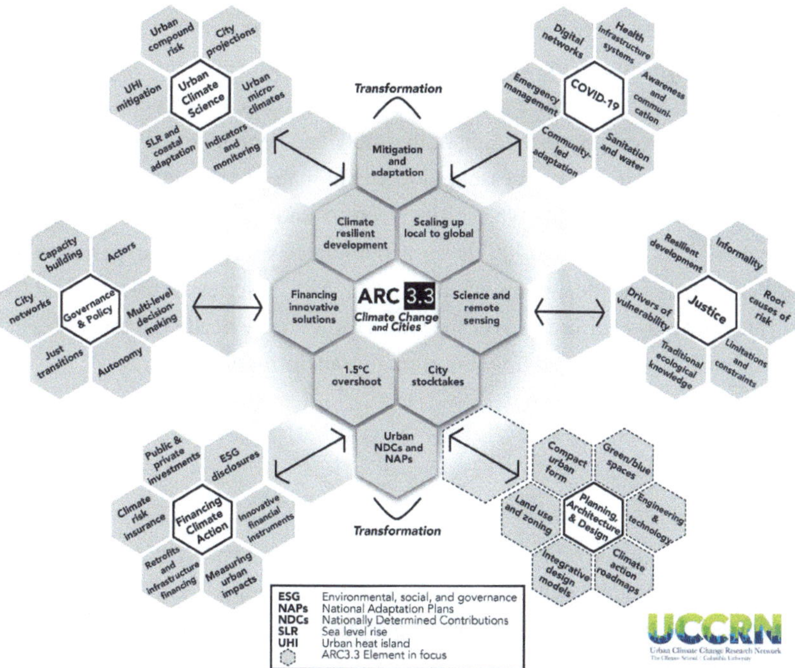

Figure 1 Cross-cutting themes associated with the overall ARC3.3 assessment and the first six Elements, with Planning, Urban Design, and Architecture highlighted in lower right. This figure is also available to view at www.cambridge.org/raven-et-al.

Element and the interlinkages to related Elements. Cross-cutting themes encompass drivers of urban function, change, and management; governance of cities across municipal, state/provincial, national, and international levels; and the role of city-level models and data.

Because the fundamental contribution of the ARC3 series is to enable a learning process for urban climate action, CCTs across the ARC3.3 Elements aim to shed light on cause-and-effect relationships and elucidate effective entry-points for interventions. This focused knowledge of urban social-ecological-technological systems can inform planners, implementers, and other city actors as they undertake ways to translate the latest science into climate action in their own urban communities.

Conclusion

We are pleased to present the *Planning, Urban Design, and Architecture for Climate Action* Element of the UCCRN *Third Assessment Report on Climate Change and Cities*. This important Element sets forth a new paradigm for the intertwined research and practice needed to inform and energize climate action in urban areas. The overarching goal is to enable the transformation of the built environment to achieve an equitable and just urban future that integrates both climate change mitigation and adaptation.

This Element is a prime example of UCCRN's dedication to an ongoing city assessment process. By providing both timely and benchmarked knowledge for cities as they grow into their essential role in climate change solutions, UCCRN comprehensively presents what cities need to know to fulfill their leadership potential. This knowledge builds on accumulated and shared experience and thus advances and deepens with time. The ARC3 Series enables this active learning process in cities all over the world, now sorely needed so that cities can indeed scale up and speed up their climate transformations.

Foreword I – Kate Orff, FASLA; Professor, Columbia University GSAPP and Columbia Climate School; Director, Urban Design Program

Cities and the climate crisis are directly linked. However, it is difficult to separate the question of climate change and cities from anything else. Cities are humanity's highest collective art form, but at the same time they contain disparate and nested systems of transportation, buildings, nature, energy, waste, and agriculture. These interlocking systems are inextricably intertwined with policy, justice, law, and design contexts, and are layered over time. In the last century, most American cities have rapidly consumed land, driven by petrochemical consumption for fossil fuel-based heating and air conditioning, and shaped by road-based development patterns. This pattern has baked CO_2 emissions into the DNA of US cities and is to some degree being replicated in the global context.

Urban climate action in cities therefore requires a holistic commitment to not only a physical understanding of how these systems are intertwined but also a commitment to addressing them through forging partnerships, coalitions, and shared initiatives. It requires planners and urban designers to learn new skills that go far beyond "making" new buildings, landscapes, and compact urban forms, to creatively "unmake" carbon-consumptive patterns and places, and to unbuild with equal creative verve tempered by a deep awareness of the social injustices that need to be addressed simultaneously.

Cities are also vibrant cultural spaces and can be catalysts for change. Unequal conditions can foster suspicion, mistrust, and the breakdown of the economy and environment, but cities also enable us to gather, discuss, organize, and do better. But where do we start, and how can we do it? This UCCRN Element on *Planning, Urban Design, and Architecture for Climate Action* moves these questions farther along and challenges us to work together to apply the best-available science to this large, complex, and intertwined global urban habitat we share, and to implement the changes we need to survive and thrive.

Foreword II – Kongjian Yu, Dean and Professor, Peking University, College of Architecture and Landscape President, Turenscape

Cities have always served as reflections of human civilization, shaped by the technologies, infrastructures, and ideologies of their time – whether agricultural, industrial, or, as is imperative today, climate positive and ecological. At the nexus of concentrated human activity and nature, cities present both the origins of contemporary challenges and the opportunities for transformative solutions. The way we design, manage, and inhabit our cities will determine not only their resilience but also the survival of humanity itself.

This Element, *Planning, Urban Design, and Architecture for Climate Action*, delves into the pivotal role of urban design in addressing climate challenges. It underscores the essential position of cities in advancing the three interconnected pillars of climate action – mitigation, adaptation, and societal transformation. These pillars converge in the urban environment, demanding a shift toward a new urban paradigm – a *deep form* – that is ecologically grounded, functionally efficient, climate positive, and culturally resonant.

Deep form transcends the superficial aesthetics of conventional design and rejects the mechanistic, anthropocentric ethos of industrial civilization. Instead, it represents an integrative approach, where cities operate as extensions of natural systems. By harvesting rainwater, regenerating biodiversity, and managing energy and materials with precision, cities can evolve from being mere habitats to becoming regenerative systems. A city embodying deep form becomes a demonstration of a sustainable civilization – one that harmonizes human and ecological needs while reflecting aesthetic and cultural values.

Planning, urban design, and architecture are crucial disciplines in shaping this future. They influence how societies interact with the environment and with one another. These disciplines must embrace their responsibility to design cities that mitigate climate risks, adapt to uncertainties, and inspire societal change. They must balance ecological intelligence, technological innovation, and cultural expression, advancing the art of survival in urban design.

This Element offers a manifesto for a climate-responsive civilization. It calls for the reimagining of cities not merely as places to live but as dynamic systems that sustain life, foster equity, and embody humanity's commitment to a sustainable and harmonious future.

Foreword III – Clara Irazabal, Director of the Urban Studies and Planning Program in the School of Architecture, Planning and Preservation at the University of Maryland

Planning, Urban Design, and Architecture for Climate Action arrives at a pivotal moment, offering a comprehensive exploration of how cities can catalyze the urgent transformation needed for a sustainable future. As humanity grapples with the escalating climate crisis, cities stand at the forefront of both challenge and opportunity. Representing over 70 percent of global greenhouse gas emissions, cities are significant contributors to climate change and critical players in its solutions, both mitigation and adaptation (Lwasa et al., 2022).

This Element's remarkable breadth encompasses disciplines from urban planning, design, and architecture to environmental justice and climate finance. Each section provides actionable insights grounded in research and enriched by case studies, underscoring the universality of challenges and solutions. The Element engages with innovative tools and methodologies and urban transformation's social and ethical imperatives, emphasizing equity and inclusivity.

Cities are more than infrastructures; they are hubs of diverse populations, cultures, innovation, and resilience. Urban leaders, planners, and residents can harness their strengths to design neighborhoods and regions that are climate-adaptive, mitigative, and just. By integrating science, policy, and practice, this Element demonstrates how cities can implement transformative initiatives that respond to local needs while contributing to global climate goals. By embracing integrated strategies that address mitigation and adaptation, cities in the Global South can lead the charge in fostering climate justice – reducing inequities, enhancing resilience, and ensuring that marginalized communities are central to planning processes. Examples in this Element illustrate how these cities can emerge as laboratories of innovative solutions, blending traditional knowledge with cutting-edge practices to address local and global challenges.

This Element is more than a grand knowledge repository; it is a clarion call for bold, collaborative, and inclusive climate action. Whether you are an educator, policymaker, practitioner, or concerned citizen, this resource should inspire and equip you in our fight against climate change. We can make cities beacons of hope and engines of sustainability. The time for action is now!

Series Editors Introduction to *Planning, Urban Design, and Architecture for Climate Action*

The ARC3.3 Element *Planning, Urban Design, and Architecture for Climate Action* presents the work of nearly fifty authors who together assess both theory and practice, as well as the multidisciplinary, cross-sectoral, and social justice issues embedded in climate change and urbanization challenges that confront cities.[†] The result is an Element that reflects the scientific, technical, operational, and social dimensions of forward-looking research and action in the fields of planning, urban design, and architecture. It is a guidebook for the transformation of cities in response to climate change that at the same time advances sustainability and equity.

Designing, planning, and building cities to address climate change mitigation and adaptation is a complex task that involves the transformation of the energy, food, water, transportation, housing, and waste disposal systems that are physically embedded in the city. But a city is also shaped by human needs for social interactions and the expression of cultural identity, including connection to nature. Social, cultural, and behavioral patterns are embedded in the structures of urban life, giving rise to lifestyles and consumption patterns. Cities can therefore also be spaces for reimagining, reinventing, and redesigning a just future under changing climate conditions.

This Element is a great "How-To" manual for accomplishing these lofty goals. It lays out how to conduct multidisciplinary and multi-stakeholder co-design of just climate interventions. It is a comprehensive assessment of methods, tools, and

---

[†] Suggested Citation: Raven, J., Leone, M. F., S. Bhaduri, C. Braneon, D. Corbett, D. Driskell, U. Eicker, J. Fernández, G. Jing, A. Hürlimann, I. Judah, M. Neuman, B. Norman, D. Pamlin, R. Chao, R. Roggema, P. Salehi, A. Shellum, A. Souza Santos, J. Towers, C. Visconti. 2025. Planning, Urban Design, and Architecture for Climate Action. In Solecki, W., M. Pathak, M. Barata, A. Barau, M. Dombrov, and C. Rosenzweig (Eds.), *Climate Change and Cities: Third Assessment Report of the Urban Climate Change Research Network*. Cambridge: Cambridge University Press.

techniques that practitioners can use to actualize and accelerate mitigation and adaptation. This duality of the Element, with its deep roots both in the research literature and everyday practice, helps elucidate real ways that the enormous task of climate capacity building in cities around the world can be accomplished.

As a concrete way to bring together the multifaceted interactions of research and practice for individual cities, the Element presents examples from UCCRN Urban Design Climate Workshops (UDCWs). These sessions bring together urban designers, urban planners, climatologists, policymakers, stakeholders, and graduate students to develop tools and methods that identify, configure, and evaluate responses to evidence-based climate challenges at the range of scales germane to cities. UDCWs co-generate just climate implementation actions that consider governmental, developmental, socioeconomic, and ecological conditions. Co-developing these perspectives enables cities to take on concrete and effective implementation of climate action.

The Element also points to where the twinned spiral of research and practice needs to proceed next. For example, the authors highlight the paucity of research on sustainable infrastructure in urban areas in low-income countries. Fostering research and practice collaborations in these and all countries is critical to the creation of a future with low carbon, resilient, and equitable cities.

Accelerating climate solutions requires research and practice communities to transition toward an open, integrated, equitable, and collaborative new model that constructively influences the very nature of design in the current moment and in the future. At its core, *Planning, Urban Design, and Architecture for Climate Action* addresses this urgent need by formulating a transformative model for design research and practice. The model enables climate action in the built environment through assessment of research, field testing, and validation of urban systems frameworks that integrate mitigation and adaptation. To enable such integrated and collaborative climate action while meeting unprecedented sustainability and social justice challenges, the Element defines and calls for new knowledge exchange and expertise, modes of engagement, technology application pathways, and multi faceted, equity-based civil society engagement.

Major Findings and Key Messages

Major Findings

1. **For urban climate actions to succeed, they must be integrated with solutions to immediate challenges facing cities and their inhabitants**. Cities around the globe face multiple stresses, with many residents beset by insecure tenure, economic turbulence, and growing inequality, as well as increasing climate

extremes. Holistic planning and design that addresses the full range of stresses, including climate, is an important new path to effective action. (Section 4)

2. **Urban services and associated co-benefits generated by climate-resilient planning and urban design represent a significant opportunity to address past and current inequalities experienced by vulnerable groups and areas**. Top-down climate action initiatives often prioritize central urban core areas and fail to benefit historically marginalized neighborhoods. Since only a very low percentage of cities have fully integrated justice and equity effectively into their climate actions, there is great potential for restorative justice through climate initiatives. (Section 6)

3. **Moving from top-down planning and implementation toward local participatory processes can help to integrate justice and equity into climate resilient development**. In the context of planning and urban design practices, collaborative processes for knowledge-sharing and co-design in multi-stakeholder contexts are key to embedding inclusive capacity-building practices, while co-producing essential knowledge components integrated into project development. A shift to inclusive, climate-sensitive development can only be realized through collaboration with many stakeholders, including local residents as well as policymakers, investors, and researchers. (Section 6)

4. **Climate policy is often developed at the city, regional, national, or even international level, but the neighborhood is a nimble scale for urban designers and planners to effectively test and deploy innovative climate-aligned strategies**. Focusing on neighborhoods, while integrating policy information across relevant urban scales, is essential for developing practical urban design and planning actions to respond to climate change (Section 5).

5. **Adaptation is occurring in fragmented and incremental ways and is lagging behind mitigation**. Few urban climate change plans consider mitigation and adaptation jointly. There is great potential to integrate adaptation and mitigation goals at building, neighborhood, city, and metropolitan scales. Mitigation and adaptation strategies include compact land use development, sustainable mobility, green and blue infrastructure, renewable energy utilization, as well as recycling and reuse of water and solid waste (Section 5).

6. **Incorporation of urban climate science is essential for the implementation and testing of innovative and practical urban design and planning climate actions**. Urban climate research provides background information, observations, and climate change projections for metropolitan, city, neighborhood, and building design scales at the mesoscale (50–500km), local scale (1–50km), and micro scale (1m–1km). The spatial scales in these two systems are correlated, enabling both communities of practice to understand each other reciprocally and to transfer knowledge effectively. (Section 8)

7. **Planning, design, and architecture that incorporate climate strategies foster co-benefits that proliferate across urban systems**. Observed social, economic, and environmental co-benefits fulfill human needs and enhance the quality of life of urban communities. Specific key co-benefits of climate actions include increased quality of public spaces and access to social services; employment and income generation from new green jobs creation; improved quality of water, soil, and air; and increased urban biodiversity. (Section 4)

8. **A transition to a more equitable, climate-integrated, collaborative, and rapidly deployable design paradigm is underway**. Elements of the new paradigm include understanding and appreciation by researchers and practitioners of overlapping interests and expertise; applying off-the-shelf digital platforms to rapidly enable and foster sharing and communications between the sectors; governments and private sector investment in applied research to advance innovation and learning on climate change in cities; integrated educational opportunities for students in professional programs learning about climate change and the built environment; rapid deployment of research findings to the marketplace; and promotion of sustainable infrastructure research in all regions of the world. (Section 2)

Key Messages

1. **Transformative urban climate adaptation and mitigation will require evolution and innovation of planning discourse, governance, and tools, as well as greater integration of design practice across spatial scales**. Systemic transformation that simultaneously encompasses metropolitan region, city, neighborhood, and building scales will motivate decision-makers, urban planners, designers, and architects to generate new practices for thinking, organizing, and acting that support climate action and justice.

2. **Urban climate justice including restoration of past harms and greenhouse gas emissions reductions need to be embedded in all development projects along with solutions for current challenges**. Urban transformation requires centering the lived experiences and priorities of marginalized communities, ensuring that climate action simultaneously addresses structural inequalities and environmental risks. Incorporating baseline and historic studies of local vulnerabilities alongside climate analyses enables development projects to align mitigation and adaptation measures with community-defined needs and goals.

3. **Enhancing participatory urban climate-resilient planning and deployment is an important priority for policymakers and practitioners alike**. In the context of planning and urban design practices, collaborative processes for knowledge sharing and co-design in multi-stakeholder contexts are key to embed inclusive capacity-building practice while co-

producing essential knowledge components integrated into project development. A shift to inclusive, climate-sensitive development can only be realized through collaboration across many stakeholders, including policymakers, investors, researchers, and residents.

4. **All projects should include both mitigation and adaptation**. Integrating mitigation and adaptation within urban planning, design, and architecture is essential to ensure cities address both the drivers and impacts of climate change. There is a need to highlight more deliberate strategies that align long-term resilience with emissions reduction efforts.

5. **More attention needs to be paid to climate action at the neighborhood scale**. The fine-grained spatial scale of a neighborhood enables local communities to be actively engaged in the process of decision-making, design, and implementation for adaptation. The neighborhood scale also provides important mitigation opportunities, such as compact neighborhood development, which decreases resource use and prevents urban sprawl, thereby reducing GHG emissions.

6. **Design professionals should proactively interact with urban climate scientists**. Collaboration between design professionals and urban climate scientists enables the testing of relevant climate change scenarios and interventions for specific urban areas. Such cross-sectoral engagement ensures that urban form and spatial strategies are informed by climate science, strengthening the potential for transformative action at all design scales.

7. **Professional planning, design, and architecture organizations need to enable and promote effective capacity building for climate action among their members**. Climate action focused on capacity building is lagging. To respond to this deficit, climate-aligned standards, training, and guidelines for practitioners should be advanced to build professional capacity. Design guidelines play an important role in capacity building by integrating and transforming multi-disciplinary scientific knowledge into intuitive and straightforward planning practice and should align development needs and local priorities.

1 Introduction and Framing

Faced with cities' urgent need for concrete climate action, planners, urban designers, and architects play a key role in developing climate action roadmaps for buildings, districts, cities, and regions across the world. The purpose of this Element, *Planning, Urban Design, and Architecture for Climate Action* in the *Third Assessment Report on Climate Change and Cities* (ARC3.3), is to motivate planning and design strategies for urban climate action that act in synergy with short- and long-term social, economic, and environmental goals and thus deliver sustainable solutions tailored to local priorities. Transforming cities for climate resilience is a cross-sectoral challenge. This Element provides concrete ways planning, design, and architecture can contribute to climate-resilient development, which comprises combined blueprints for climate change adaptation and reduction of greenhouse gas emissions, while supporting a socially just and sustainable future for all (IPCC AR6).

In UCCRN's *Second Assessment Report on Climate Change and Cities* (ARC3.2), the "Urban Planning and Urban Design" chapter defined a set of evidence-based urban climate factors aimed at integrating mitigation and adaptation actions (Raven et al., 2018; Rosenzweig et al., 2018). This Element builds on that chapter to develop tools and methods to assess – and ultimately configure – climate-resilient urban transformation by bridging climate research and climate action. The Element assesses these strategies from region to city and from neighborhood to building scales, with a primary focus on urban neighborhoods.

To integrate climate mitigation and adaptation, this Element assesses research, models and simulations, field testing, and validation of urban system strategies, built environment configurations, and policy frameworks (Figure 2). It draws from the work of policymakers, scientists, designers and planning practitioners, private-sector actors, non-governmental organizations, and community experts. Its goal is to enable integrated climate action in cities through expansion of the capacity and agency of planners, urban designers, and architects to create roadmaps for urban transformation.

A shift to inclusive, climate-sensitive development can only be realized through collaboration across and among stakeholders, including governmental civil servants, investors, researchers, and residents; and through alignment and coordination of action across scales, from the neighborhood to the citywide, regional, and national scales. Urban practitioners are both trained and positioned to facilitate this cross-sectoral, integrated change.

Because cities are currently responsible for an estimated 70 percent of global greenhouse gas emissions and with urban systems exceeding their ecological capacity,[5] they require massive built environment retrofits and widespread adoption of sustainable lifestyles (Tan et al., 2021; IPCC, 2022). Radical and replicable shifts from food, energy, water, transport, waste, buildings and natural systems to efficient and integrated urban systems are crucial to achieve climate-aligned mandates and ambitions.

Local characteristics of the built environment drive the intensity and impact of climate change in cities (Oke at al., 2017). Local climate hazards are aggravated by the profound transformations that have been caused by the growth of settlements and the effects of concentrated activities in urbanized areas. Urban systems and

---

[5] Ecological capacity is an ecosystem's ability to produce biological materials used by people and to absorb waste material generated by humans (GEMET, 2021). See www.eionet.europa.eu/gemet/en/concept/15125.

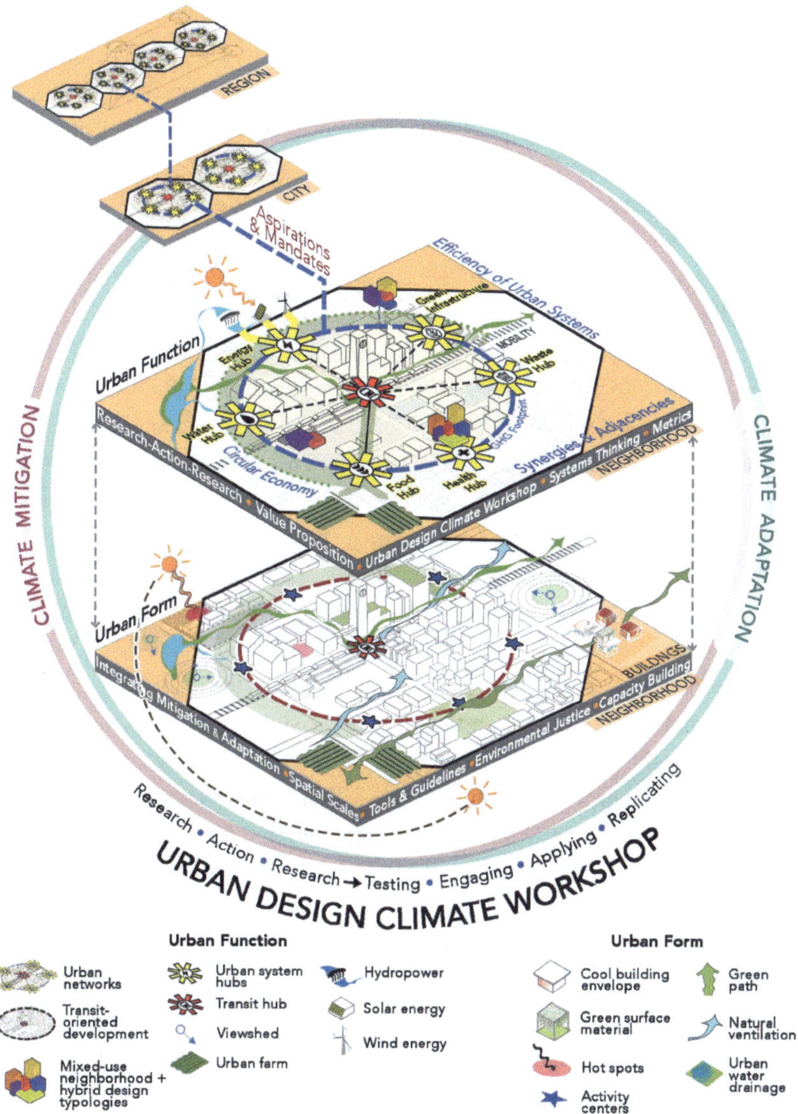

Figure 2 Framework for climate-aligned urban design (Raven, J. 2025).

human behavior affect both local climate vulnerabilities and GHG emissions. They shape local microclimatic conditions and require varied response capacities.

At the same time, through organizations such as the Global Covenant of Mayors for Climate and Energy (GCoM), cities are consolidating their role at the forefront of collective action and policy innovation to tackle the climate emergency (Rosenzweig et al., 2025). Many cities have set ambitious climate targets and are working to accelerate the transition by acting as replicable laboratories for testing and applying innovative mitigation and adaptation strategies, multi-stakeholder partnerships, and financing models (GCoM, 2022). The 2022 IPCC Sixth Assessment Report (AR6) estimates that compact and resource-efficient cities could reduce global GHG emissions by a quarter by 2050 compared to a business-as-usual scenario (IPCC et al., 2022).

At present, the predominant way in which cities are being developed neglects the comprehensive action necessary to meet the Paris Agreement target of limiting warming to 1.5°C (UNFCCC, 2015).[6] Science indicates achieving this target would require the reduction of global greenhouse gas (GHG) emissions by 43 percent by 2030 and 84 percent by 2050 (IPCC, 2022). These targeted reductions are commonly translated to 50 percent by 2030 and to net zero by 2050 (as recognized in the, e.g., Glasgow Climate Pact [UNFCCC, 2021]). Simultaneously, cities must also adapt to the changes that cannot be avoided. Therefore, the status quo must be disrupted if the extent of climate change is to be limited, and for cities to be well adapted and sustained under these new climate conditions.

The goal is to develop the knowledge foundation needed to meet the unprecedented urban challenges posed by climate change. City case studies are embedded throughout the Element to provide insights from actions undertaken by cities from a wide range of geographies. These cities are testbeds for the approaches, methods, and tools described in this Element. Using case studies, practitioners can learn how other cities are adapting their climate actions to local circumstances such as demographics, geography, climate, culture, and financial capacity.

The *Planning, Urban Design, and Architecture for Climate Action* Element thematically connects with several ARC3.3 Elements on compound climate risks, resilient development, smart-energy grids, and green and blue infrastructure (Figure 3). Local climate data must be used to ensure effective planning and urban design; infusing science-based design with community insights can reduce vulnerabilities; incorporation of traditional ecological knowledge will advance just adaptation; and cost-effective infrastructure can be usefully developed that integrates mitigation and adaptation.

This Element is divided into eight sections that present key concepts associated with the paradigm shift in planning, urban design, and architecture theory and practice required for climate transformation in cities.

- *Research Informing Practice, Practice Informing Research* highlights the dynamic paradigm between climate action and research. Persistent barriers to

---

[6] Evidence from multiple sources including the IPCC, global datasets, reports from national and local agencies, and peer-reviewed literature confirms that climate change is occurring more rapidly and ubiquitously than previously projected, with 1.5°C overshoot likely (Schleussner et al., 2024; Rosenzweig, et al., 2025).

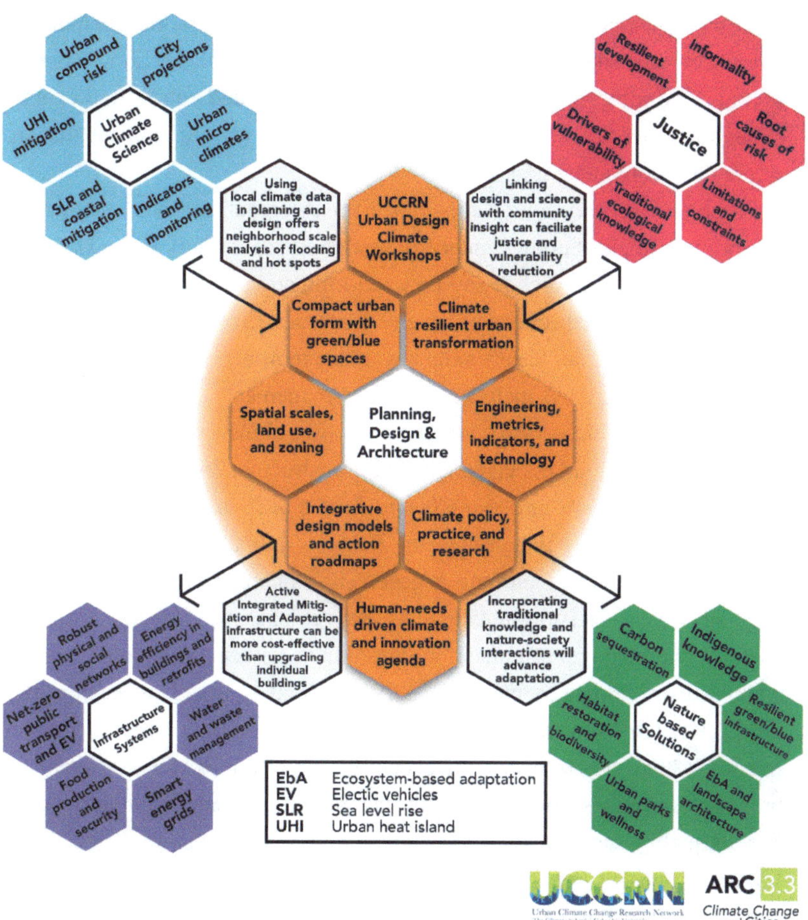

Figure 3 Cross-cutting themes linking ARC3.3 Planning, Urban Design, and Architecture to other Elements.

confronting climate challenges led by local practitioners and stakeholders can illuminate urgent research gaps. In response, these challenges shape a pragmatic, problem-solving research agenda and in turn, can be applied and tested by stakeholders and practitioners to evaluate positive climate outcomes.

- *Urban Transformation through an Expanded Climate and Innovation Agenda* reflects on the need to involve urban practitioners, innovators, communities, and decision-makers in order to promote rapid climate-resilient development with a focus on human needs and societal well-being. Architects, urban designers, and planners are called to stimulate transformative change through disruptive, dynamic, non-linear, and systemic visions that reflect the impact of innovation.
- *Climate Resilient Urban Transformation* analyzes the return on urban transformation investments and introduces how pragmatic climate policies benefit

from synergies between social aspirations and urban systems. Included are cross-sectoral benefits of urban transformations to public health, economic savings, and job creation shaped by climate justice and equity, and examinations of "carrot and stick" incentives by public, private, and institutional actors that motivate action.

- *Integrating Mitigation and Adaptation* focuses on actions that synergize the two primary goals of climate intervention through the lens of applied research. In place of the usual dichotomy between mitigation and adaptation that can lead to uncoordinated actions or adverse outcomes, the section explores synergistic activities that are ideally suited to urban environments and better allocate limited resources. This includes climate-management activities designed to reduce GHG emissions while producing win-win climate benefits through heat stress reduction, flood management, and infrastructure resilience.

- *Embedding Environmental and Climate Justice in Planning and Design* integrates this societal challenge into the fabric of climate-driven planning and urban design processes through engagement on community priorities, resilience strategies, decision-making, and project phasing. Climate action can present economic opportunities for cities, but these benefits should be sure to address distributional, contextual, and procedural equity. This section highlights climate-driven transitional justice, job training, and circular economy initiatives that enable access to economic opportunities for disadvantaged groups. It describes how differential vulnerabilities, such as climate hazards and air quality, can be addressed by urban design and planning.

- *Capacity Building for Urban Decision Makers and Practitioners* describes current research on stakeholder processes and explores how community experts, policymakers, and practitioners can strengthen knowledge sharing, co-design, and co-evaluation practices. Effective local climate action draws from a set of shared parameters including equity in funding, inter-agency collaboration, and recognition of common priorities. Collaboration with the full range of stakeholders involved in climate-resilient transformation of cities can help balance top-down and bottom-up perspectives.

- *Metrics, Performance Indicators, and Tools* presents assessment methods and state-of-the-art approaches for adaptation and mitigation, ranging from digital and modeling tools to intuitive climate-based design guidelines to direct stakeholder engagement practices. The questions are how to assess performance, evaluate success, and leverage climate research for replication and up-scaling. Spatial scales are analyzed in terms of required cross-sectoral expertise, tools, and methods, from buildings to local neighborhoods to metropolitan regions.

- *Urban Design Climate Workshops (UDCWs)* describes an innovative action-driven framework, developed to confront climate change in cities and identify opportunities for cross-sectoral refinement (Raven, 2021). It provides examples from multiple UDCWs with urban design practitioners, researchers, policy-makers, stakeholders, and graduate students8 working side by side. The UCCRN team has partnered with city officials, communities, and practitioners to support local roadmaps on climate action in cities across the globe by jointly confronting city-specific challenges and applying methods and tools to support the climate-resilient planning and design.

- *Conclusions and Research Gaps* summarize vital information on planning, urban design, and architecture for climate action and recommend future work.

2 Research Informing Practice, Practice Informing Research

Our changing climate is posing an unprecedented set of challenges. The built environment is the result of an innumerable number of multifaceted decisions informed by science, engineering, architecture, urban design, planning, and many other disciplines. Cities have long faced threats from war, disease, and pollution. Concerns regarding the human-affected urban microclimate have been part of urban planning but became particularly critical during the Industrial Revolution. The urban heat island (UHI) phenomenon was first described and measured in the early 1800s, both anthropogenic sources of heat and effects of urban morphology, built surfaces, and vegetative cover (Howard, 1818). (See Additional Resources for further historical information related to the built environment in cities.)

Deep collaboration and sustained sharing of knowledge among architects, planners, engineers, and scientists are required to effectively and promptly address the evolving threats of climate change, while rapidly reducing GHG emissions from the built environment. This section explores this critical need for knowledge production and sharing to enable a more resilient and sustainable built environment.

2.1 International Climate Policy Processes and the Built Environment

The Intergovernmental Panel on Climate Change (IPCC) has cited the built environment as one of the most important, economical, and practical sites for immediate and substantial GHG emissions reductions (Lucon et al., 2014; Cabeza et al., 2022). The technology solutions to achieve a reduction in energy use in the built environment by mid-century exist today: these include high-performance and low-cost insulation materials, units glazed with solar control films and gases, high-efficiency heating and cooling systems, energy-efficient appliances and equipment, and smart and digital control systems (GlobalABC/IEA/UNEP, 2020).

However, obstacles that currently hinder widespread adoption of such low/ zero-carbon and adaptive solutions in the built environment are weak building codes, ineffective zoning that does not prevent urban sprawl, lack of financing solutions for higher upfront investments, imperfect information, lack of awareness of available technologies, and industry fragmentation. There are also vested interests in maintaining 'business-as-usual' practices that utilize the same construction methods as before, thus limiting choices for consumers.

The 26th United Nations Framework Convention on Climate Change Conference of the Parties (COP26), held in Glasgow in 2021, had positive outcomes that have created momentum for better embedding climate action into the built environment sector. Directly relevant is the commitment "urging parties to further integrate adaptation into local, national and regional planning (s8 and s51)." Furthermore, more than twenty-four countries and a group of leading automobile manufacturers committed to ending fossil fuel vehicles by 2040. The outcome of COP26 confirmed that a major global challenge is building and retrofitting more climate-sensitive cities. This challenge for the built environment can only be tackled successfully by a much stronger interactive connection between knowledge and practice, and thus between research and practice.

COP27, held in Sharm-el-Sheik in November 2022, further expanded the focus on buildings as a critical industrial sector to achieve decarbonization. It highlighted the embodied carbon of materials and technologies used in the construction sector, in addition to carbon emissions from building operation. The establishment of a Loss and Damage Fund now underway is expected to accelerate the development of solutions simultaneously targeting climate mitigation and adaptation by investing in resilient infrastructure, promoting sustainable design, and enhancing urban capacity-building among stakeholders.

At COP28 in Dubai, UN-Habitat promoted the role of cities in addressing climate change and launched the call for climate action through transformative urban planning, together with leading organizations including C40 Cities, Urban Partners, United Cities and Local Governments (UCLG), and enhancing the Our City Plans toolbox through climate change and infrastructure modules. At COP29 in Baku, more than 160 stakeholders, including countries and cities, endorsed the Multisectoral Actions Pathways (MAP) Declaration for Resilient and Healthy Cities, with an emphasis on advanced planning and design for urban sustainable transport, green construction, urban agriculture, and nature-based solutions.

2.2 Research and Practice

Research communities and practitioners are generally served by conferences, journals, and academic and professional societies that foster intradisciplinary interactions, collaborations, and communications. However, *intra*disciplinary communication and collaboration in research and practice communities have not resulted in ease of *inter*disciplinary communication and collaboration to the extent needed for rapid development of effective climate actions. Communication and collaboration among fields begin to diverge at the university level when students choose to specialize in planning, urban design, architecture, engineering, or other fields. Many professional educational programs separate these disciplines upon matriculation into one field or the other. As students graduate and become certified by the relevant professional organizations, the disciplinary separation is locked in.

The separation is even greater between the built environment professional fields and the physical, biophysical, and social sciences. The sciences generally are taught in academic units (departments, colleges, schools) that are separate from those specializing in planning, urban design, architecture, and engineering.

Furthermore, research communities remain substantially siloed from practicing communities and vice versa. The complex and evolving challenges of climate change action – embedded in both mitigation and adaptation – are not easily separated into discrete issues that can be effectively tackled by one discipline. Many evolving climate threats – for example, sea level rise, extreme heat, and increasing heavy downbursts – require multiple research and practicing communities to come together to understand the fundamental science of the phenomenon, develop science-based solutions, and deploy resources and professional knowledge from practice in the real world (Rosenthal et al., 2007).

Fortunately, a significant portion of research focused on the built environment is generally considered applied, where research inquiries are guided by agendas in fields such as engineering and urban climate science informed by opportunities for actualized deployment in buildings and cities. These research agendas for the built environment are now fundamentally influenced by the challenges of climate change.

Partly because of the applied nature of this work, researchers today can connect to practice in several ways: through networks; joint conferences and special issue publications focused on climate change; integrated foundational and professional educational opportunities (see Box 1); innovation activities, including startups; and consultations and engagements with industry and building professionals. However, there is a critical need for broader, deeper, and more sustained communications and interactions between researchers and practitioners, both those who are ready to engage with climate change and those who are already addressing its challenges.

Today, research and practice communities can dramatically strengthen their interactions to accelerate the emergence and deployment of solutions to climate change (Fallman et al., 2020). This requires building new bridges and strengthening existing ones to enhance two-way interactions and communications between the priorities and insights of the research community and those of practitioners.

Box 1 Building Bridges between Research and Practice:
Pedagogy and Paradigm

The education of design practitioners represents one of the most important opportunities to challenge paradigms that isolate knowledge production in the academy from professional know-how in the field. Climate change is transforming education as a matter of pedagogy and curriculum, resulting in a fundamental rethinking of the relationship between research and practice. Sustainability as a paradigm requires broad social engagement, collaborative and transdisciplinary teaching structures, and the capacity to integrate and seek out knowledge well beyond isolated professional disciplines or "best practice" models. Pedagogically, the challenge is teaching students the skills they need to excel in the marketplace, while giving them the knowledge they need to remake that very marketplace (Towers, 2005).

Three recent examples exemplify this work at the macroscale of universities and the microscale of graduate and undergraduate courses. In May 2021, the Massachusetts Institute of Technology (MIT) released Fast Forward, an updated "Climate Action Plan" to their 2015 prior commitment. By leveraging research, innovation, and education, they align the mission of their institution with the challenge facing humanity at large: "find[ing] affordable, equitable ways to bring every sector of the global economy to net-zero carbon emissions no later than 2050" (MIT, 2021).

Based on similar institutional commitments at the Parsons School of Design, a course-by-course evaluation of the undergraduate architecture program curriculum aligns the urgency of the climate crisis with educating the next generation of designers (Hagan, 2022). For architecture this means exploring, researching, and understanding the systems of extraction, production, manufacturing, assembly, use and life cycle of materials, and the new paradigm of of circularity (cradle to cradle) versus linearity (cradle to grave).

In 2012, the New York Institute of Technology graduate urban design program aligned the focus of its curriculum with the multi-year development of Urban Design Climate Workshops (UDCWs).[7] UDCWs encompass urban climate science, policy, legislation and governance, multi-scale planning and design, engineering, ecology, and social sciences, engaging students in real-world urban planning experiences.

---

[7] Supported by the National Science Foundation. In 2022–2024, the Erasmus+ funded project UCCRN_edu expanded this collaboration by experimenting with innovative multi- and interdisciplinary educational pathways and integrating higher education curricula courses.

Architects and planners worldwide understand that climate change is a present and growing challenge for the built environment. However, specific agendas that drive research in the social and biophysical sciences focused on the climate are not well known and not broadly understood by individuals and organizations in professional practice (Hosey, 2017). This is the case in most regions of the world, and it is thus reasonable to state that the typical architect is not deeply knowledgeable about the science of climate change.

Similarly, scientists and engineers working in specialized and highly technical fields are not generally familiar with the priorities of practicing communities of planners, urban designers, architects, and engineers. Specific points of common interests can motivate more effective communication and deeper collaboration between research and practice.

The need for creating bridges between research and practice is being answered by boundary organizations and city networks. The role of nonprofit, advocacy, think tank, and other types of organizations in advancing the deployment of research into practice and facilitating connections between practitioners and researchers is crucially important. Organizations such as the Urban Climate Change Research Network, World Green Building Council, Local Governments for Sustainability, C40 Cities Climate Leadership Group, and the World Economic Forum develop practical climate-informed research to bridge the science-to-practice gap.

As planning and urban design in cities involves a wide range of stakeholders, research and practice will be bridged only if they arise from integrative approaches that employ *co-design* (Webb et al., 2018). UCCRN's UDCWs bring research into practice by exploring how urban design and planning can drive impactful climate action in urban areas through a co-designed and bottom-up approach (see Section 9). Embedding transformative approaches into the planning process, such as through UDCWs, enables cities and stakeholders to move away from models and prototypes to actionable solutions.

2.3 Design in Transition

For both research and practice, solving fundamental questions of how cities change and how they can increase their resilience to climate shocks demands new forms of inquiry and practice. Cities are socio-ecological-technological systems (SETS) operating in specific geographic locations with multiple intersecting social, geographical, historical, economic, and ecological factors (Figure 4) (McPhearson et al., 2022; Chester et al., 2023).

Enabling integrated climate action in cities requires expanding the agency of urban planning, urban design, architecture, engineering, and construction. Evidence-based, cross-sectoral design and planning processes require development and rezoning scenarios that include climate mitigation and adaptation. The process

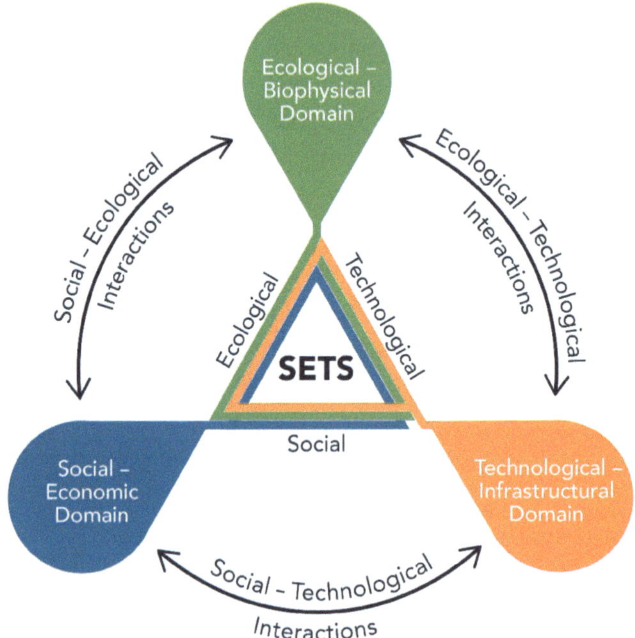

Figure 4 The SETS framework acknowledges the interactions and interdependencies among the social–cultural–economic–governance systems (Social), climate–biophysical–ecological systems (Ecological), and technological-engineered infrastructural systems (Technological) that drive urban patterns and processes (McPhearson et al., 2022).

includes visioning, stakeholder exchange, and solutions testing, while critically engaging with established planning paradigms. Desired outcomes include zero-carbon urban mitigation strategies and adaptation interventions for biodiverse, flood-resilient, and green cities with cleaner air, lower summer temperatures, ubiquitous urban encounters with nature, and enhanced civic life.

Climate action also requires integration across multiple scales – telescoping through metropolitan region, city, neighborhood, and building lenses. Embedding climate action in urban areas needs to occur at each level to ensure it is incorporated into strategic planning and urban design, local area and neighborhood development, and building construction, each supporting the others.

Research informing practice and practice informing research may involve different approaches at each of these spatial scales. Some focus on more governmental aspects, others more partnerships with the private sector, and others with academia. New platforms for connections are increasingly required to support multi-stakeholder exchanges – for example smart digital platforms – to enable greater sharing of knowledge, co-design and co-production of innovative planning, and urban design solutions. The digital city approach, i.e. a city that utilizes information technologies and web-based platforms to manage its infrastructure, services, and citizen interactions, provides greater opportunity for collaboration when accompanied by appropriate ethical considerations such as privacy (Norman, 2018; Yigitcanlar et al., 2021). These tools can provide important spatial platforms for more efficient integrated decision-making, such as the Smart City Initiatives in Rio de Janeiro (Calzada, 2021).

Problem setting and solution formulation in the context of climate change requires new methodologies of engagement and knowledge production to generate the system level change that is called for. "Transition Design" is a new methodology that has emerged from the scholarship of design practice in response to climate change (Norman & Stappers, 2016; Irwin, 2019). Transition Design draws on approaches from the social sciences to understand the historical and cultural roots of problems and places, stakeholder concerns, co-design, and collaboration at the center of the problem-solving process. Traditional design approaches (characterized by linear processes and decontextualized problem frames, whose objective was the swift realization of predictable and profitable solutions) are inadequate for addressing this class of problems. (See Additional Resources, Figure 1, for the Transition Design Framework.)

2.4 Accelerated Climate Solutions

The urgent need for substantial climate action in the built environment this decade demands a concerted effort to connect the latest science with the best practice (IPCC, 2018). Accelerating climate solutions requires that research and practice communities transition toward an open, integrated, and collaborative new model that constructively influences the nature of design. Climate justice, sustainable infrastructure in developing countries, and co-design are three key elements of this new model.

Climate justice is a key element of this transition. Research and practice can prioritize climate justice into their distinct modes of inquiry and add climate justice as a measure of the implications of a new technology to vulnerable communities. In setting priorities, architects, planners, building professionals, owners, and investors of the built environment can incorporate progress toward climate justice as a fundamental measure of success.

Additionally, there is a paucity of research on *sustainable infrastructure in developing countries* (Thome et al., 2016). Fostering important research/practice collaborations in developing countries is critical to a future of low carbon and equitable cities (Costello & Zumla, 2000). Effective solutions will also arise from integrative approaches in a continuous and self-reinforcing manner from a systems approach that employs *co-design* (Webb et al., 2018).

Key considerations for research and practice in advancing climate change actions in cities include:

- Understanding and appreciation by academics and practitioners of the valuable contribution of different perspectives
- Smart digital platforms that enable and foster better sharing and communications between sectors
- Investment by governments and the private sector in applied research that can advance innovation and learning of urban climate change
- Integrated educational opportunities for professional students learning about climate change and the built environment
- Rapid movement of research findings to the marketplace
- Promotion of sustainable infrastructure research especially in developing regions

3 An Expanded Urban Climate and Innovation

Although there has been progress in adaptation governance and planning, with ~170 countries including adaptation in their climate policies and planning processes, there remains a significant gap between the actions required and those taking place (Béné et al., 2018; Torabi et al., 2018). Existing governance and planning systems and their processes and components (e.g., discourses, structures, tools, and practices) in the Global North and South are intrinsically challenged by their multi-functionality and trade-offs between functions across temporal and spatial scales (Asadzadeh et al., 2023; Bush & Doyon, 2019).

Although incorporating coping and incremental adaptation measures in established governance processes and planning has shown some ability to manage the challenges associated with climate change, emerging evidence shows their incapability to drive the transition toward more just and sustainable cities. Relying on only conservative coping and reformative incremental approaches in shaping already-in-operation governance and planning systems can sharpen existing mismatches, generate opposite effects, exacerbate existing or new vulnerabilities, and cause maladaptation (Anguelovski et al., 2016; Ziervogel et al., 2016).

Many adaptation responses are fragmented, incremental, sector-specific, and unequally distributed. These result in differing levels of implementation, significant gaps in adaptation and mitigation, and, in some cases, maladaptation, which can severely impact marginalized groups. Thus, transformative mechanisms are needed to fundamentally change existing governance and planning systems. Transformative adaptation urges decision-makers and planners to generate new mechanisms for thinking, organizing, and action that support resilience and incorporate them into planning discourses, governance structures, tools, and practices (Hölscher et al., 2019). Such systematic transformation can potentially reorient planning discourses, reorganize governance structures, innovate planning tools, and expand planning practices at various scales and locations.

Initial climate work focused on reducing emissions from existing systems rather than on solutions needed for global sustainability. This section argues that this gap is due to a lack of representation and direction, and a disconnect between human needs and innovation that motivates a limited and short-term approach to addressing complex challenges. Specifically, instead of addressing root causes, most current practices focus on symptoms that arise from existing systems. This shortcoming is mainly due to a reductionist approach to carbon emissions and ignoring the complex interrelationships from a systems perspective where human needs are intertwined with sustainable development. Repositioning human needs at the center means asking fundamental questions about existing systems and motivations.

There is a wealth of solutions to the many challenges that already exist – whether currently connected to climate or not. In the prevailing approach to climate action, there is pressure for cities and metropolitan regions to arrive at the table with a predefined climate-focused solution. This is exclusionary and against the imperative that all actors and sectors must be involved to achieve the 1.5°C threshold (Rosenzweig et al., 2025; Schleussner et al., 2024). Therefore, now more than ever, it is essential to "meet cities where they are." This means identifying successful solutions from across the globe, assessing whether they are climate focused or could be adapted to have a climate focus in the future, and exporting and scaling them.

Transformative change is disruptive and systemic, occurs at relatively large scales, requires multiple actors, and involves reconfiguring technology, economy, institutions, and society, including paradigms, goals, and values (Visseren-Hamakers et al., 2021). The scope of transformation is, therefore, on discourses (dominant narratives, persuasions, theories), structures (arrangements, organization, administrations), tools (strategies, plans, projects), and practices (norms, rules, incentives). Systems transformation refers to changing a system's underlying structures, patterns, and dynamics to achieve a desired outcome.

In response to the challenge of operationalizing transformative change, this section explores the connection between innovation, technology, and human needs, proposing an Expanded Climate and Innovation Agenda (ECIA). This agenda proposes a dynamic approach to addressing the climate crisis that puts human needs at the center, leverages the technologies of the fourth industrial revolution, and encourages the identification and export of solutions from cities and regions around the globe. Under this agenda, a new innovation ecosystem is required that incorporates all actors, including incubators and start-ups, to contribute to collective action toward addressing climate change. The section starts by outlining the ECIA before unpacking the human needs approach. It then moves towards implementation by identifying capacities needed for the ECIA to come to fruition before concluding with a call to action.

3.1 A Human Needs-Driven Agenda

If the goal of urban and spatial planning is to develop "good" cities and make them better places to live, it is essential to define these concepts to understand how planning could best achieve them. Defining the notion of a "good" city has been a topic of debate by urban scholars for decades, with notable scholars, including Lefebvre (1996) and Friedmann (2000), contributing to the discourse. Utopian visions of cities are not intended to be a blueprint but rather to establish a foundation of values informed by contextual factors that offer transformative and creative alternatives to current practices.

Harvey (1973) called for "genuinely humanizing urbanism," and Lefebvre (1968) for "the right to the city." Amin (2006) argues that a "good" city is one built upon "an ethic of care incorporating principles of social justice, equality, and mutuality." He summarizes this idea as "contemporary urban solidarity." How to achieve this solidarity hinges upon the capacity of institutions of governance to engage with local complexity. The "good" city contains robust climate action plans that balance the economic activity necessary to sustain livelihoods and urban processes while innovating new methods to nurture and enhance existing social and environmental systems. This requires fundamental transformation at the systemic level to redefine development and how success is measured (see Case Study 1). Kaika (2017) argues that humans are "living indicators" of the quality of urban environments and that learning from the experiences of individuals through established platforms for inclusive engagement in decision-making processes is essential.

One illustrative example of a city iteratively following on this approach is Naples, Italy, which is building local climate mitigation and adaptation strategies and identifying synergies with community needs, with a focus on urban redevelopment of peripheral areas and the application of UDCW methods and tools since 2018. Local communities and stakeholders are engaged through collaborative mapping exercises and parametric design tools to assess climate mitigation and

CASE STUDY 1 LEVERAGING THE WATER–ENERGY–FOOD NEXUS: SYNERGISTIC OPPORTUNITIES FOR CLIMATE CHANGE MITIGATION AND ADAPTATION IN LEH[8]

Daphne Gondhalekar

Leh, located in the Indian Himalayas, is a rapidly transforming small city facing serious water scarcity and wastewater management issues. As the population grows and tourism is steadily rising, groundwater extraction and wastewater seepage is increasing. In 2009, the Ladakh Autonomous Hill Development Council commissioned an international consulting company to address this challenge. They designed a centralized sewerage system with a conventional energy-intensive sewage treatment plant at the foot of Leh. In 2014, construction began, however, as project costs increased over the years, the system could not be constructed as planned and, today, only half of Leh can be connected.

Since 2012, the Nexus Research Lab at the Technical University of Munich, has been collaborating with Leh through a human needs-driven agenda to address the city's continuing water stress through the design of a Water–Energy–Food (WEF) Nexus approach for replication in the area of Leh that is not connected to the centralized sewerage system. Stakeholder workshops and consultations have been held in Leh, involving local NGOs, government, academia, civil society, and spiritual leaders.

The WEF Nexus approach acknowledges that planning water, energy, food, and other relevant sectors in conjunction, rather than in "silos" as is conventionally done, can support more efficient use of natural resources, thereby lowering GHG emissions (Hoff, 2011).

Wastewater is generally an untapped resource that can serve as a valuable alternative source of water, particularly in water-stressed regions (UNESCO, 2017). Decentralized water reclamation with integrated resource recovery can yield several inherent benefits in Leh, such as supporting water conservation, as decentralized systems are smaller and require less water to flush. Water can be reused locally, for example, for toilet flushing, groundwater recharge, and agricultural irrigation. Lifting and treating less water also implies energy conservation, and organic fertilizer can be recovered from fecal sludge, which can improve soil fertility. Decentralized water reclamation with resource recovery can support water, energy, and food security in Leh as well as be a more manageable and affordable option for the local population.

---

[8] See extended version of case study at https://uccrn.ei.columbia.edu/case-studies.

CASE STUDY 1 (cont.)

In 2023, Leh's local government put together a consortium to secure funding to implement the Nexus pilot project, starting in 2024. Using a Nexus approach, cities can become not only consumers but producers of valuable resources to regenerate entire city regions (Gondhalekar & Ramsauer, 2017).

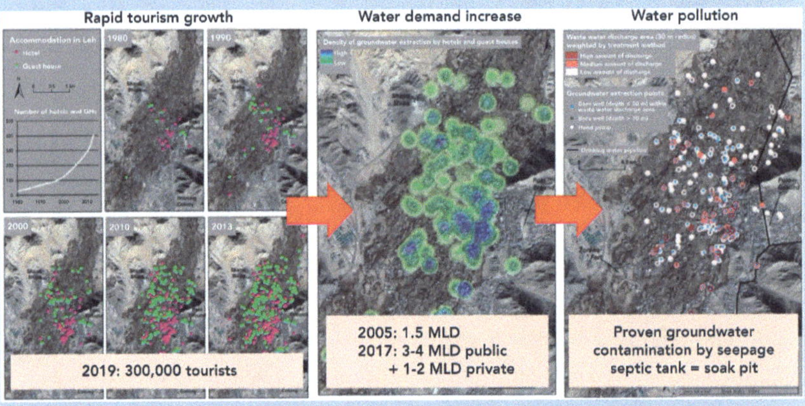

Figure 5 Groundwater extraction in Leh quadrupled from 2005 to 2017 (million liters daily [MLD]), and accrual of wastewater pose threats to drinking water quantity and quality.
(Source: Gondhalekar, 2023)

adaptation potential of different design scenarios. The outcome of these workshops resulted in a set of user-friendly analysis tools which showed the effects of heat and flood risk on buildings' layout and morphology, surface materials, vegetation and public spaces. This work expanded in 2024 through the development of the Naples Sustainable Energy and Climate Action Plan (SECAP) for GCoM. (See the additional resources for more details on the Naples UDCW).

In response to the notion of a good city, a human need-driven agenda contains two parts. First, all humans are included, both current and future generations. This means the solutions must work in a world where 8 billion people today and 11 billion in the future will live – with the majority residing in cities. The resource efficiency and cost efficiency required for such a world necessitate a new approach to everything from climate and innovation policy to global collaboration and city planning. Second, all human needs must be met. This includes the basic human needs required for survival, that is, nutrition/health, shelter, security, and social and cultural needs that allow individuals and cultures to flourish. In parallel, ecosystem balance is considered an essential component of human well-being and an ethical agent with an intrinsic value (see the expanding concept of "rights" by Nash, 1989).

A focus on human needs provides an opportunity to ask and challenge various stakeholders, including companies, how they are making life better for people. This is essential for cities and regions as they are responsible for providing and enabling fundamental services for their citizens. Most companies only provide a part of the solutions needed, and many are even undermining human needs by creating artificial needs, manipulating markets, and by selling more than is needed. Food and goods value chains in general have a direct impact on the physical dimension of cities, profoundly impacting on available spaces for public services, housing quality, energy consumption, pollution, and waste generation.

Achieving climate-resilient urban transformation thus requires considering the impact of innovation and emerging economic sectors in cities and society as a whole. To effectively address this gap in the context of urban transformations, it is essential to reflect on the impacts that technological (product or process) innovations have on the spatial, functional, and environmental components of cities. New types of spaces and functions – such as coworking, maker spaces, circular hubs, technological and creative offices – reflect the strategies of both industrial entrepreneurs who propose innovative development models, and of the subjects who control land use, economic policies, and urban infrastructures (Zukin, 2020). A good example concerns key infrastructures linked to digital economy models, rooted in recurring transformative land use patterns, which include executive offices in central urban areas, data centers in isolated areas, and fulfillment and logistics hubs that often contribute to urban sprawl.

On the other hand, big tech and e-commerce companies are increasingly demanding green building standards for their critical infrastructure, and the size of their investments in cities can become an important asset to drive transformation towards more equitable, sustainable, and climate-resilient goals (Grainger, 2024). Nevertheless, making these investments synergized with community and environmental priorities requires going beyond current approaches that are mostly driven by corporate sustainability policies towards a more systematic dialogue between companies, city officials, planners/designers, and citizens (Temple, 2024).

This is particularly relevant for projects of new business/innovation districts, where issues related to rising house prices and the loss of integrity of multigenerational neighborhoods are observed, and of new critical infrastructures (e.g., logistic hubs, data centers) negatively impacting urban sprawl, soil consumption, and private car usage. Careful consideration should be paid to systemic impacts of such transformations and the induced system dynamics in a life-cycle perspective. These projects should better tackle issues such as embodied carbon of materials (e.g., steel and glass) and spatially and thermally inefficient form, rather than only balancing these factors with increased renewable energy production. Anticipating the need for adaptive reuse of office buildings, avoiding the risk of new monofunctional areas, and using existing transit-oriented and mixed-use urban design are also areas for careful consideration (Raven et al., 2018). (See Additional Resources for information on scaling up of sustainable innovative solutions.)

3.2 Planning and Urban Design Capacities for Implementation

Most city planning related to climate change has so far been based on static problem assumption, a compliance approach focused on reducing GHG emissions and where the city is seen in isolation. An expanded climate and innovation agenda also includes a dynamic solution approach where the city prioritizes human needs in a way that is compatible with a world where the global population can live flourishing lives in the next decades, as well as the role of the city as a solution provider and community enabler (Table 1). For example, the New York Climate Exchange, located on Governors Island in New York City, is a recently developed global center that harnesses the combined strengths of education, research, workforce development, policy creation, and public programming to drive climate action on local, national, and international scales (New York Climate Exchange, 2021).

This gives rise to two fundamental questions central to urban planning:

1) *Facilitating Flourishing Lives*: How can urban planning ensure the well-being of citizens while aligning with the context of natural boundaries (Rockström et al., 2009)? This challenge holds particular significance for the Global North, given its resource-intensive existing systems that necessitate transformation.

2) *City-Led Solution Export*: How can cities position themselves to actively export essential solutions that address global needs and goals?

Table 1 outlines five capacities that can help cities implement an Expanded Climate and Innovation Agenda by addressing the two questions, with a focus on planning and urban design methods and tools. (See Additional Resources, Figure 14, for the role of the innovation quadrant and start-ups.)

Table 1 Key planning and urban design capacities for implementing an expanded climate and innovation agenda.

Capacity	Motivation and cross-sectoral implications	Application in planning and urban design practice
Scan, map, and benchmark climate and innovation strategies.	Current climate strategies are often compliance and reduction driven, overlooking local innovations and incubators that address human needs and export solutions.	Introduce self-assessment tools to map climate and innovation strategies, and benchmark city strategies against indicators such as human needs. Engage urban stake-holders in discussions.
Use a full iceberg approach to move beyond tip-of-the-iceberg responses.	Urban governance models struggle with complex challenges; moving from reactive to agenda-setting innovation requires a "full iceberg approach" to build local institutions' capacity to address root causes and complexities.	Assess planning and urban design solutions with social, environmental, and economic co-benefits using simulation, benchmarking tools, and stakeholder co-evaluation of multi-sectoral impacts. Explore "full iceberg" implications of planning and urban design strategies for generating new data, reshaping systems, enhance well-being, and align local actions with global goals.

Table 1 (cont.)

Capacity	Motivation and cross-sectoral implications	Application in planning and urban design practice
Deploy *11 billion innovation quadrants*[a] to guide strategies towards globally sustainable system solutions.	An 11 billion innovation quadrant focusing on GHG mitigation and cost-efficient land use can ensure that global solutions are developed and enabled in cities.	City planners can build capacity for international climate innovation and test benefits of initiatives by farming strategies and solutions within the quadrant.
Gather and demand human need-based data.	Expanding the climate and innovation agenda requires data focused on human needs, not just optimization of existing systems, to challenge current infrastructure models and support smarter, more sustainable solutions.	City planning should prioritize data that supports both optimization and sustainable solutions, including downscaled climate data, socio-economic data including crowd-sourced community insights, and environmental data from sensors. Planners and urban designers could increasingly adopt information-driven tools to provide evidence-based assessments and simulation of expected effects of climate-resilient development solutions for buildings and open spaces.

Table 1 (cont.)

Capacity	Motivation and cross-sectoral implications	Application in planning and urban design practice
Integrate start-ups as both local and international solution providers.	Start-ups and incubators, which focus on societal needs, should be considered stakeholders included in city planning.	Collaboration between city planners and incubators can connect start-ups and practitioners and cluster initiatives to address city- and community-specific needs through global-local networking.

[a] The 11 billion Innovation Quadrants is a framework that categorizes and visualizes potential innovations that includes addressing the needs of a projected global population of 11 billion, environmental sustainability, and climate change impacts (Mission Innovation, 2023).

Notes: (See ICLEI and Mission Innovation/RISE tools and benchmarking and assessment tools for cities, companies, and practitioners)

3.3 Systems Thinking, Multi-Disciplinarity, and Multi-Stakeholder Visioning

Addressing critical challenges from the nexus of urbanization and climate change cannot be accomplished without a system-based approach, including social science, ecology, and technology. Strategies to plan and develop a more just and resilient future require cities to face multiple climate risks that overlap in space and time. Strategies to build resilience to one event may decrease resilience to another event happening at the same time and place (Elmqvist et al., 2019).

Systems knowledge in the context of climate change and cities relates to understanding the current state of critical urban systems (e.g., energy, transport, material flows, water supply and management, food, and waste). Net-zero carbon, climate-resilient infrastructure, designed and delivered through a systems approach, can directly benefit people's health, well-being, livelihoods and the environment. For example, urban density is supported by well-designed transport systems in tandem with effective urban planning. Urban density creates economies of agglomeration, reduces commuting times, and supports social capital. New infrastructure projects and retrofits of existing infrastructure can create significant employment opportunities while driving economic gains (Grainger, 2022).

The generation of *target knowledge* and *transformation knowledge* calls for creative imagination of new system concepts and an articulation of pathways that can link unsustainable present states of urban systems to envisioned sustainable future states (Gaziulusoy & Ryan, 2017). In the envisioning process, multidisciplinary stakeholders are brought together to collect ideas and formulate a joint vision that can take the form of qualitative or quantitative goals and targets, combining data gathering, analysis, and interpretation.

The complexity and diversity of cities and related vulnerabilities inevitably lead to a plurality of perspectives, values, visions, and knowledge systems that define what cities are or should be. *Transdisciplinary urban systems science* and practice can mobilize a diverse array of knowledge creators to develop resilient futures. Scientific data, quantitative risk analysis, and computational models are important to such transdisciplinary science, but not sufficient. Intangible, non-material flows and dynamics of urban systems, such as how different people experience risks or how they connect and interact with other groups in the city to build social capital, are challenging to measure and model (McPhearson et al., 2022). This raises an urgent need for an "urban systems science" that includes multiple forms of knowledge collaboratively produced to define a legitimate and transformative urban climate action process (Romero-Lankao et al., 2018).

Co-production of knowledge includes processes that iteratively unite ways of knowing and acting – including ideas, norms, practices, and discourses – leading to mutual reinforcement and reciprocal transformation of societal outcomes (Wyborn et al., 2019). It embraces multiple ambitions and engages multiple actors (researchers, decision-makers, and citizens) to produce new knowledge and new ways of integrating knowledge into decision-making and action, ultimately producing new outcomes. Decision-making cannot be conducted in a silo but rather an iterative participatory process.

This points to a strong need to advance knowledge and policies that are currently available with continuous interaction between multiple city actors, building the climate response together from the beginning. For this to happen, both science and policy institutions can help to advance diverse participatory initiatives that are not a threat to their power or something to be controlled but rather an opportunity for learning (Galende-Sánchez & Sorman, 2021). Imagining and co-producing shared positive visions is the basis for developing transformative plans, policies, and actions to drive

Table 2 Conditions for success in local collaborations (elaborated from Adame, 2021).

Experiencing local context	Consider Indigenous groups, local universities, commercial sectors, and non-governmental and governmental organizations.
Meeting people in the field	Collect and analyze local data and discover communalities.
Sharing knowledge	Offer collaboration and opportunities to engage. Give credit for data sharing and analyses (e.g., authorship on reports and journal papers).
Engaging continuously	Involve a variety of urban stakeholders by sharing information and asking for feedback early and often.
Valuing local actors input	Recognize local actors and acknowledge their added value in results and outcomes.

toward building a longer-term, more just, resilient, and sustainable world (McPhearson et al., 2021).

Collaboration to achieve transformative solutions to climate change in cities requires that experts from a range of disciplines and representatives from the city's communities and government interact substantively. These interactions depend on a set of critical factors for success. Adame (2021) points out the following five elements for success in stakeholder interactions: 1. Experience of local context; 2. Meeting people in the field; 3. Enabling knowledge sharing; 4. Engaging continuously, 5. Valuing local actors' input (see Table 2).

As opposed to 'helicopter research,' which is far removed from local contexts and jumps to quick conclusions, recognition of the diversity in understanding among all participants improves the research itself. In this sense, local communities are an important resource for the conceptualization of local and regional solutions for complex and unprecedented climate change (Roggema et al., 2017).

While there is no shortage of institutional drivers, standards, guidance, and training for practitioners to draw upon to enhance capacity engagement, the procedural mechanisms for effective collaboration are lagging behind intentions, declarations, commitments, strategies, and action plans.

4 Climate-Resilient Urban Transformation

Cities and urban neighborhoods around the globe face multiple challenges, with many residents stressed by insecure tenure, economic turbulence, growing inequity, and – most recently – a global health pandemic.[9] Climate change exacerbates many of these challenges and poses an existential threat, but one that is less tangible and immediate to most people than issues such as paying next month's rent, taking care of a sick loved one, feeding a family, or replacing a lost job. The success of urban climate solutions often depends on how well they are integrated with solutions to these immediate and pressing challenges. It is important for climate action to create and connect with opportunities for new livelihoods, community wealth building, greater equity, and healthier, more inclusive cities.

Effective urban climate solutions address both climate change and other pressing challenges, such as poverty, inequality, and lack of opportunity. Urban planners and designers are now embracing the broader value proposition of climate strategies, that is, the economic, social, and environmental co-benefits. Climate solutions that also address near- and long-term urban development needs are more likely to gain public and private buy-in, political support, and necessary resource commitments, thereby accelerating, broadening, and deepening their impact.

This section identifies how built environment responses to climate change can generate other tangible near- and long-term benefits and be more likely to gain political support and community buy-in. It identifies equitable methods for defining and measuring value, and ways in which value propositions impact planning decisions, built environment investment priorities, and outcomes.

4.1 Generating Value with Climate Strategies

For many urban residents, daily life is tenuous. Lack of housing, rising rents, low-paying jobs, poverty, crime, environmental hazards, and limited access to basic services (education, health, safe and reliable water, and sanitation) create significant stress and require immediate time and attention. While severe weather events and chronic stresses from a warming planet are bringing long-term climate threats to the forefront of policymaking, they remain often removed from the daily concerns of many city residents.

The widespread equity imbalance in urban conditions – both globally and locally – has resulted from decades of resource-extraction practices and inequitable consumption patterns. In the global economic order of the 2020s, wealth imbalances have never been greater (Hasell et al., 2023). Wealthy nations are consuming more resources and emitting more GHGs than ever before, while poor nations are left to fend for themselves and bear the brunt of increasingly

---

[9] See ARC3.3 Elements, *Justice for Resilient Development in Climate-Stressed Cities*, and *Learning from COVID-19 for Climate-Ready Urban Transformation*. www.cambridge.org/core/publications/elements/elements-in-climate-change-and-cities

disruptive climate trends. The necessary shift away from carbon-intensive urban practices is not just a shift away from polluting fuel sources: it represents a profound opportunity to address the systemic imbalances that created the climate crisis in the first place.

Climate action in cities has demonstrated a potential pathway to addressing development needs, including job creation, improved public health, social inclusion, and increased accessibility (see Case Study 2) (Gouldson et al. 2018). To realize

CASE STUDY 2 A CLIMATE RESILIENT URBAN LANDSCAPE IN WESTERN PARKLAND CITY, SYDNEY[10]

Rob Roggema

The metropolitan area of Sydney is expected to grow from approximately five to over seven million people in the next twenty years. Locations of new housing and required infrastructure are issues that need to be addressed at the larger scale. The Greater Sydney Commission has developed a long-term spatial strategy, dividing the total area into three complementary cities. In this strategy, the majority of new housing is planned in the so-called third city of the Western Sydney Parklands. This area will be home to over one million new residents over the next ten to twenty-five years and requires a large-scale urban plan in which future resilience is the condition for newly built homes.

Western Sydney is part of an inland climate zone that sits within the mountainous surrounds and therefore, captures heat, dust, and is relatively hidden from sea breezes, causing accelerated climate impacts. Extreme heat leads to bushfires and droughts in the summer months, and severe flooding during the rainy season. In this context, urban growth in this area implicitly increases the vulnerability of its residents, as more and more people will be exposed.

To improve climate resilience for future populations, city planners have placed the landscape at the heart of future regional planning through a series of six nature-based mapping stages. In Stage 1, the current water network is captured, with the flow of main water courses based on the local topography, determining the ecological gradients and pathways of discharging water flows through the landscape. In Stage 2, alongside the streams, a riparian zone was planned in which wet forests would create the ideal conditions for sensitive ecologies, increasing biodiversity and capturing surpluses of water, preventing flooding. Stage 3 entailed the introduction of an ecological grid

---

[10] See extended version of case study at https://uccrn.ei.columbia.edu/case-studies.

forming the frame for structural ecological connections, linking wet and dry, and nutrient-rich and -poor parts of the landscape to increase ecological connectivity and exchange, enhancing eco-capacity. Stage 4 introduces plantations of timber forests in the inner parts of the landscape to produce building materials for homes. The final two stages will be designed to host agroforestry and free-range livestock farming.

This planning strategy shows the way landscapes can be put first in developing a new urban precinct. The way ecology, water, and soil systems form the underlying layers of the landscape, hence defining the landscape tissue for land use activities, is an approach that can be applied in every region in the world, to create a resilient and climate-adaptive urban future.

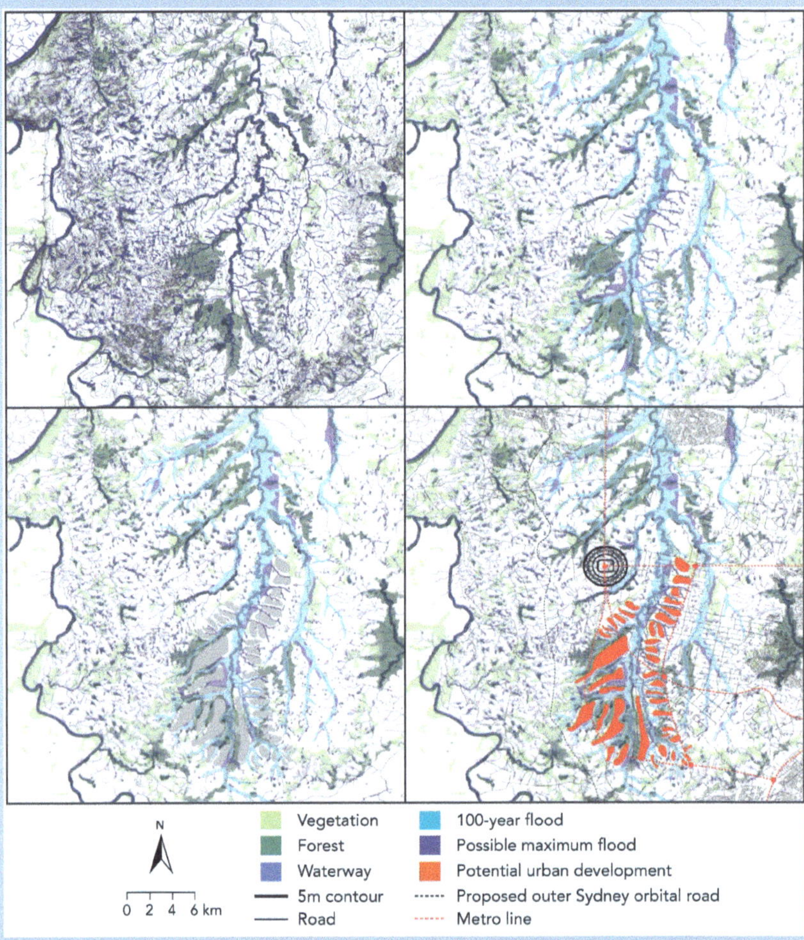

Figure 6 Topographical regional maps showing flood risk and potential areas for development in Western Parkland City, Sydney.
(Source: Roggema, 2023)

these outcomes for all stakeholders, municipal governments need to make explicit commitments to prioritize climate investment in ways that benefit their most vulnerable populations and to address historic inequities. In absence of such commitments, the benefits of sustainable design and planning have accrued disproportionately to the wealthy (Mahendra et al., 2021). Without unprecedented effort to protect the urban poor and vulnerable from the steadily increasing impacts of climate change, city economies and populations will be placed at extreme risk (IPCC, 2022).

4.2 Adding Value through Co-Benefits

The potential economic benefits of addressing climate change in cities are increasingly clear. The Coalition for Urban Transition found that an annual investment of US $1.8 trillion in existing technologies and strategies could cut 90 percent of global GHG emissions, generate annual returns of $2.8 trillion in 2030 from energy and material cost savings, and create a net present value of $16.6 trillion by 2050 (Coalition for Urban Transitions, 2019; Gouldson et al., 2018). The cost of inaction is also well studied (UNEP, 2023).

The World Bank estimates that by 2030, climate change could drive an additional 132 million people into poverty, and the annual climate-related spending for low and middle income countries other than China is estimated at $783 billion between now and 2030 (World Bank, 2022). Urgent infrastructure interventions include adaptation of energy systems, creation of coastal-defense systems, and heat and wildfire mitigation measures. As the scale of change required to respond to climate change continues to mount, so, too, does the potential return on investment for cities.[11]

Climate-action economics are also increasingly favorable in the private sector, and real estate investments are becoming tied to strong sustainable development outcomes. As of 2023, the Global Real Estate Sustainability Benchmark (GRESB), an Environmental, Social, and Governance (ESG) reporting tool for real estate, included approximately 1,800 real estate investment fund members representing $8.4 trillion in combined asset value. Approximately 50 percent of funds are based in Europe, 27 percent in the Americas, 15 percent in Asia, 7 percent in Oceania, and 1 percent in Africa (GRESB, 2023). As of 2021, the UN Principles for Responsible Investment reporting had more than $121 trillion in assets under management (UNEP-PRI, 2021).

---

[11] See ARC3.3 Element, *Financing Urban Transitions to Climate Neutrality and Increased Resilience in Cities, www.cambridge.org/core/publications/elements/elements-in-climate-change-and-cities*

While the correlation between climate action and economic outcomes is strengthening, the broader value proposition for communities requires further investigation. It is not guaranteed that high-performing real estate assets will yield true environmental or social benefits (Kempeneer et al., 2021). Nor are positive returns on investment for municipalities and private real estate investors automatically shared equitably across local communities. Similarly, urban development decisions based on GHG emissions targets alone often do not provide near-term value to residents and could potentially exacerbate or perpetuate existing inequities. The impacts of climate change are projected to increase in all urban regions in the coming decade, even as mitigation efforts ramp up (IPCC, 2022).

Urban interventions can be designed to provide near-term value beyond emissions reductions to meet city needs and garner support, purposefully integrating equity criteria in their design and delivery. The UN proposes a broad definition of urban development value as the "totality of economic, environmental, social, and intangible outcomes that have the potential to improve quality of life for residents in meaningful and tangible ways" (UN-Habitat, 2020). To achieve this, cities need to prioritize climate actions that have the potential of delivering multiple benefits.

Another framework for urban planning climate initiatives illustrates the concept of Integrated Climate Action. Effective strategies to achieve net-zero cities are intentional to advance the three core goals of sustainable development (economic growth, social inclusion, and environmental protection) and related co-benefits while reducing GHG emissions (Figure 7). They are enabled by complementary policy, action, and investment at different scales. Value proposition for climate resilient urban design and planning is strengthened when clearly emphasizing links between mitigation/adaptation goals and local socio-economic priorities.

In this Element, the co-benefits of climate action are referred to as "the development and implementation of policies and strategies that simultaneously contribute to tackling climate change whilst addressing local environmental and developmental problems" (UN; Puppim de Oliveira, 2013). When well defined and understood amongst stakeholders, co-benefits can help increase stakeholder support for urban projects, enable mobilization and funding within city agencies (see Box 2), and increase the speed and impact of climate actions (Floater et al., 2016; Bain et al., 2016). Cities citing the co-benefits of their climate action reported 2.5 times more climate actions than cities that did not (Bachra et al., 2020).

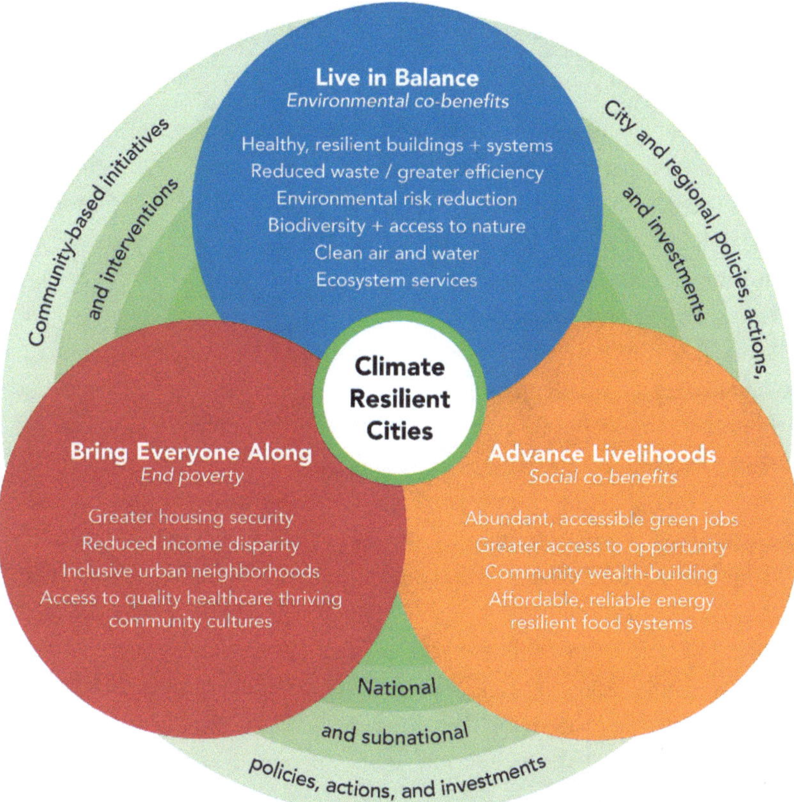

Figure 7 Integrated climate action framework: Value proposition for climate resilient urban development. (Driskell, D. 2025)

4.2.1 Measurement and Quantification of Co-Benefits

Once potential co-benefit opportunities are identified, a balanced means of measuring the value of these benefits is required (Ficklin et al., 2017). Many studies primarily focus on the *potential* of climate development strategies to generate co-benefits rather than systematically evaluating and reporting results (Suckall et al., 2020). Measuring social value generation and human well-being indicators related to climate action is difficult. Challenges in quantifying climate co-benefits include lack of alignment on definitions and measurement practices, intangibility of co-benefits, difficulty in attributing system change to specific interventions, and insufficient cost and benefit data (Lou et al., 2022). Sectoral and governmental silos present additional barriers by preventing the aggregation of data and expertise (Suckall et al., 2020).

Green buildings present an increasingly better investment opportunity than business-as-usual structures due to reduced operating costs and higher and longer-term value retention (Likhacheva Sokolowski et al., 2019). However, methods for sharing costs and economic benefits for both residents and investors often require non-traditional funding and investment approaches. One such example exists in the South African Affordable Housing Real Estate Development Fund, managed by the IHS Towers Company.

In order to encourage green development techniques while meeting required returns for investors, the International Finance Corporation (IFC) contributed an equity investment of US$10 million with funds from the Global Environmental Facility (GEF) to cover the marginal costs of sustainable design, as well as technical support, in order to achieve the EDGE (Equity, Diversity, and Gender Equality) certification (Likhacheva Sokolowski et al., 2019). These interventions reduced utility bills for affordable housing residents equivalent to one month's rent annually, allowing developers to demonstrate competitiveness in a cost-sensitive market.

In addition, the incremental costs of EDGE certification have decreased over time thanks to technological advances and stricter building codes. Consequently, most developers participating in the fund now pursue green buildings due to its lower costs and competitive nature (Likhacheva Sokolowski et al., 2019).

Emerging data collection and processing techniques are helping to address these challenges. The proliferation of sensor technology and associated data tools such as machine learning and artificial intelligence can support evidence-based urban development decision-making for mitigation and adaptation (Leal Filho et al., 2022; Yagoub et al., 2022). Data and technology advancements enable urban planners to correlate urban form with human cognition along with citizen, stakeholder, and expert input. In particular, public datasets and machine learning techniques are increasingly used to support built environment research on the relationship between urban infrastructure and quality of life outcomes (Biljecki, 2021).

However, while data are essential to quantifying the value proposition of urban climate strategies, machine learning and artificial intelligence-driven models and methods should be used with caution. Urban practitioners need to interrogate both the data and the results to vet potential sociopolitical consequences (Duarte, 2020), and professional design knowledge can be linked to urban design data processing, simulation, and recommendation techniques (Hughes et al., 2019). Additionally, data collection and processing tools can be shared with a variety of urban practitioners with limited urban science capacity, while maintaining an acceptable level of rigor. (See Additional Resources for information on the methods of evaluating co-benefits.)

4.3 Responding to Local Development Needs

Climate change is intangible and appears intractable to many urban residents. Climate-oriented planning and design that do not demonstrate clear progress in areas of higher priority for stakeholders have ultimately failed to yield both climate impact and investment returns (Jennings et al., 2020). Furthermore, such projects slow progress and dampen political will for climate-oriented development by calling into question the value of climate action. Identifying and communicating linked co-benefits encourages both public and private buy-in on climate strategies, independent of perception of climate change urgency or political ideology (Bain, 2016).

In 2021, the World Economic Forum global survey found that the top public priorities for the seventeen United Nations Sustainable Development Goals (SDGs) were "No Poverty," "Zero Hunger," and "Good Health and Wellbeing," while "Climate Action" ranked thirteenth on the list (Ipsos, 2021). However, the relationship between climate action, health, and poverty are both direct (such as in the case of cardiopulmonary diseases associated with local air pollution) and indirect (as in the case of food insecurity). In any city, urban practitioners can map interdependencies between local climate and socioeconomic issues to determine local development priorities that address both, thereby selecting climate strategies that respond to the most pressing stakeholder priorities.

Design for equitable access to co-benefits of climate projects is critical. Many top-down sustainability planning initiatives have prioritized central urban areas and assumed a spatial diffusion of the benefits to all parts of the city. They have often failed to benefit historically marginalized neighborhoods (Mahmoudi, 2019). Cycling infrastructure can reduce emissions and local air pollution while improving health, safety, and quality of life, promoting job access and reducing transportation inequality.

However, these benefits are only realized if infrastructure is built consciously to address them, and to do so equitably across the city. For example, only 23 percent of New Yorkers have access to the city's bike share system, and this same population is proportionally whiter and wealthier than the remaining 77 percent who have no access (Wachsmuth et al., 2019). Similar disparities have occurred in flood prevention planning that favors wealthier urban populations and exacerbates historic environmental injustices, inequitable access to renewable energy, and other climate-oriented investments (Liao et al., 2019; Sovacool et al., 2022).

4.4 Translating the Value Proposition to New Contexts

Urban practitioners rely heavily on precedent for informing planning and design decisions and often refer to successful case studies that demonstrate clear potential value for their city or community. However, appropriate translation of "best practice" strategies requires moving away from a business-as-usual approach and a consideration of factors that led to project success (including the political, social, behavioral, environmental, economic, technological, and geographical context) in the local project context (see Case Study 3). Without this, even the most well-intentioned climate planning decisions may not provide local value or respond to community needs. Without stakeholder engagement, climate measures may fail to realize their mitigation potential and broader development goals (Gouldson et al., 2018).

CASE STUDY 3 MAKING SUSTAINABLE, AFFORDABLE HOUSING A REALITY IN MELBOURNE

The IPCC and United Nations have identified the development of sustainable, accessible housing solutions as one of the most pressing challenges for cities in the climate transition (Habitat for Humanity, 2021; IPCC, 2022). Urban planners and designers are uniquely positioned to provide housing solutions that achieve the triple bottom line of affordability, social justice, and environmental sustainability. However, these multilayered value propositions cannot be delivered using a business-as-usual approach (Moore & Doyon, 2018).

This is exhibited by the approach of Nightingale Housing Pty. Ltd., a not-for-profit founded in Australia in 2016 to help develop and support a sustainable, affordable housing model created through iterative project learning by architect Jeremy McLeod of Breathe Architecture. In conjunction with local architects, the Nightingale goal is to provide higher density housing that properly, and equally, addresses the triple bottom line of social justice, sustainability, and affordability. The concept was first demonstrated with the development of The Commons, completed in 2013, a five-story, twenty-four-unit apartment building located in Melbourne. This development demonstrated what can paradoxically be gained – specifically, increased social interaction, as well as cost and environmental savings – through strategic elimination and reduction. The architects reduced car parking in favor of bike parking and communal car share, eliminated second restrooms and laundry in units in favor of shared facilities, and eliminated the need for AC and associated costs through passive design and renewable energy installation (Moore & Doyon, 2018).

The second development, Nightingale 1, was completed in 2017 in Melbourne. The experience and outcomes from the Commons were solidified into a set of principles for affordability, transparency, sustainability, deliberative design, and community contribution. Nightingale 1 further improved affordability by capping project profits at 15 percent and introducing a covenant that prevents the homes from selling at higher than the average price rise of the neighborhood (Moore & Doyon, 2018).

To scale the design and learnings quickly, architects can apply for Nightingale licenses to develop similar projects and receive access to the intellectual property and the waiting list of potential residents. As of 2023, there are fifteen completed Nightingale developments and seven under construction, each led by different coalitions of architects. Such developments have spread outside of Victoria to other parts of Australia and New Zealand.

Figure 8 Left: The Commons, Melbourne, with increased bike lanes, open frontage, and green wall (Source: Tom Ross, 2013).
Right: The façade of Nightingale 1, made of recycled brick. (Source: Peter Clarke, 2017).

One example is the failure of Seattle's bike share program Pronto! in 2017. While docked bike share systems have been successfully implemented in many other global cities (from Hangzhou to Paris to New York), Pronto! bikes were decommissioned after three years due to low ridership. While the attributing factors were many and nuanced (including lack of private funding and political will), improperly addressed stakeholder needs were also found to contribute. Station placement did not adequately consider hilly topography and rainy weather, commuting and tourism patterns, and existing public transportation, which ultimately limited the ridership of both members and non-members (Sun et al., 2018).

Another example is the Huangbaiyu Project in China's Liaoning province, a redevelopment project envisioned to be the "world's first eco-city." Homes constructed through this project were ultimately unaffordable to residents and remained unoccupied by local farmers who did not have enough space in the new yards to farm. Some of the houses were built with garages, though the villagers did not have cars (de Jong et al., 2016).

Research and evidence on context-specific drivers of co-benefits is limited (Gouldson et al., 2018). Much of the research has been done at an international and national level, with a small portion of research conducted at the city level (Deng et al., 2017). In particular, there is lack of co-benefits research in Oceania, South America, and Africa, which often have different development needs than those of regions in the Global North (Deng et al., 2017). Without this evidence, local urban practitioners can evaluate the viability of solutions proven in other contexts for their own project locations, in consultation with communities (Gouldson et al., 2018). This can lead to scaling of successful climate projects.

5 Integrating Mitigation and Adaptation

There is widespread recognition that adaptation will not be sustainable without significant mitigation efforts, and likewise, communities will be unable to mitigate effectively if they are focusing solely on continued adaptation to climate change impacts (Hürlimann et al., 2021a; IPCC, 2022). By emphasizing the role of "mitigation co-benefits resulting from Parties' adaptation actions" in achieving the global warming limitation goals, the Paris Agreement highlighted the opportunity for cities to prioritize strategies that integrate mitigation and adaptation actions in response to climate change (UN, 2015). This is to ensure that global warming is limited to 1.5°C and that the extent of adaptation effort required is minimized (Hürlimann et al., 2021b).

The integration of adaptation and mitigation planning and actions is critical to ensure that these are mutually reinforcing (i.e., to realize synergistic efficiencies), to maximize the impact of limited city resources, and to minimize any potential conflicts that could lead either to maladaptation or malmitigation (Grafakos et al., 2019; Santos et al., 2021). Maladaptation, actions that reduce short-term risks but increase long-term vulnerability, and malmitigation, actions that reduce emissions in the short-term but could lock in technology or practices that limit more efficient future mitigation measures, are potential pitfalls within climate interventions (Magnan, 2014; Santos, 2021).

An example of maladaptation can be found in the case of Fiji, where the construction of seawalls made coastal communities more prone to flooding, as they prevented stormwater drainage (Piggott-McKellar et al., 2020). The promotion of biofuels can be framed as malmitigation. While biofuels may cut emissions in the short-term, it risks locking in technology that undermines the adoption of more efficient fuel sources, with potential to also lead to biodiversity loss and food insecurity (Santos, 2021).

5.1 Defining Integration of Mitigation and Adaptation and Relationship to Climate Resilient Development

Integration of these two paradigms – adaptation and mitigation – recognizes the need for urban systems to both dramatically reduce emissions while simultaneously being resilient, adaptive, and transformative in the face of climate change. More explicitly, Hürlimann and colleagues (2021a) define "climate change transformation" as "the explicit integration of adaptation and mitigation actions, to achieve a new regime of 1.5 degrees of warming above pre-industrial levels by 2100, and to be well adapted to it," setting tangible mitigation and adaptation benchmarks.

Integration of mitigation and adaptation speaks to a systems-based approach to addressing urban climate change, understanding the complex interrelations that connect climate, ecosystems, biodiversity, and people (IPCC, 2023). At the foundation of this approach is the understanding that critical urban services that support communities – such as energy, mobility, materials, waste, water, and food – have been made available at the price of biodiversity loss, unsustainable consumption of natural resources, ecosystem degradation, and economic inequalities.

The IPCC WGII in AR6 introduced the concept of Climate Resilient Development as an integrated adaptation and mitigation framework that combines "strategies to deal with climate risks (adaptation) with actions to reduce GHG emissions (mitigation) which result in improvements for nature's and people's well-being" (IPCC, 2022).

In Figure 9, climate resilient development pathways show varying levels of integration of adaptation, mitigation, and sustainable development (the bigger the size, the more integration). In turn, the graph shows the consequences of lower levels of action.

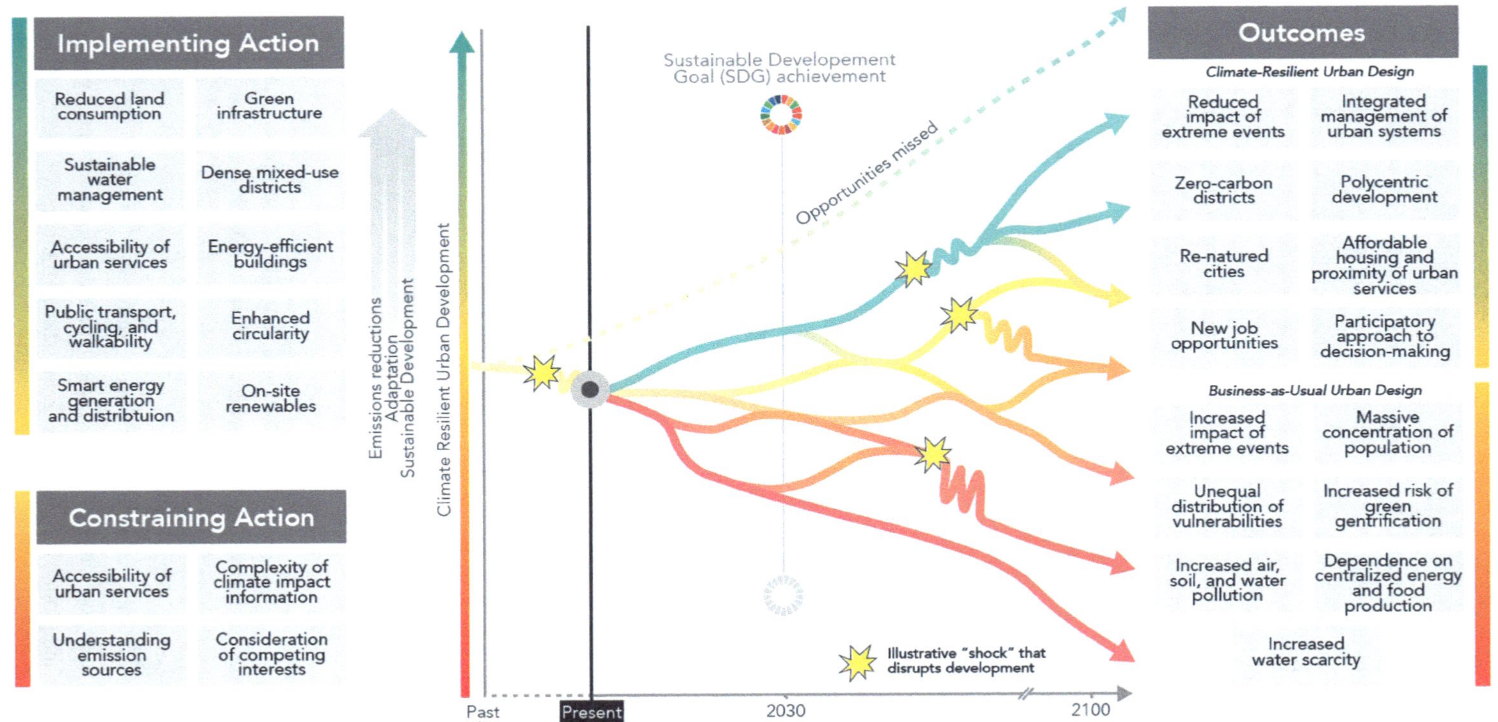

Figure 9 Climate-resilient urban development pathways (Elaborated from IPCC, 2022; Singh & Chudasama, 2021).

5.2 Role of Cities

Cities are responsible for more than 70 percent of global GHG emissions (IEA, 2021) but also highly vulnerable to the effects of climate change. Interactions between climate hazards, infrastructure systems, growing urban populations, diverse cultures, real-estate development, and economic activities in cities can exacerbate climate change and disaster impacts (Chang et al., 2018; Rosenzweig et al., 2018). Cities that substantially reduce GHG emissions while simultaneously ensuring that their land use and infrastructure adapts to a changing climate are better positioned to remain livable and economically robust in the years ahead.

The Paris Agreement brought hope that initiatives through Nationally Determined Contributions (NDCs) would help combat climate change, but urban actions appear notably missing.[12] Nationally Determined Contributions, which stem from the 2015 Paris Agreement, are climate action plans to reduce emissions and adapt to climate change. Each country that has joined the Paris Agreement develops an NDC and updates it every five years (UNFCCC, 2015). While developed countries are better prepared to provide direct assistance and support services towards adaptation and mitigation activities, developing countries need financial and technical support to implement more robust mitigation-adaptation measures (Shammin et al., 2022). This is especially true for cities.

The next decade is essential to achieving urban decarbonization through mitigation strategies for different sectors of the built environment such as mobility, waste and resource management, buildings, and energy systems (IPCC, 2023). A wide range of stakeholders needs to be included in these strategies (Norman, 2018). Mitigation initiatives are most effective when they reduce, rather than increase, risks related to climate change hazards, while urban adaptation initiatives work best when aligned with decarbonization and mitigation objectives. An example is how sustainable infrastructure can support transport systems (e.g., integrating trees and vegetative areas along tramways, pedestrian paths, and cycling routes) to offer co-benefits such as improved access to green space, boosted public/active mobility, increased biodiversity, and reduced air-soil-water pollution (see Case Study 4) (Sharifi, 2021).

Cities are ideal testbeds for innovative technology and policy development and implementation. They are shaping adaptation interventions that confront climate change at the much-needed ground level. Mitigation and adaptation strategies can drive innovative urban management to concurrently meet sustainability goals and natural hazard risk assessments (Scorza & Santopietro, 2021). Focused and central

---

[12] See ARC3.3 Element on *Governance, Enabling Policy Environments, and Just Transitions,* www.cambridge.org/core/publications/elements/elements-in-climate-change-and-cities

administrative oversight of relatively small-scale projects bodes well for cities to lead in coupled mitigation/adaptation strategies.

Lee and colleagues (2020) suggest that adaptation is positively influenced in cities with mitigation action policies through monitoring systems, rather than mere mitigation commitments. This is evident in the case of cities worldwide that are signatories of GCoM and commit to updating their Sustainable Energy Action Plans (SEAPs) into Sustainable Energy and Climate Action Plans (SECAPs), which use quantitative indicators to concurrently monitor the benefits of implemented projects and measure for mitigation and adaptation. The study also finds that national mandates are important drivers of local adoption of adaptation policies. In Copenhagen, the city's adaptation strategy aligns with national laws on energy efficiency in the building sector, enhancing regulations for retrofitting buildings to improve energy use while also incorporating disaster risk preparedness.

Cities need to be well equipped and prepared in the face of increasing climate change-induced disasters (Sharifi & Yamagata, 2015). Preparedness should include governance, finance, and climate policy aspects. Implementation of goals and policies in practice is largely determined by the robustness of institutions (e.g., strategic plans, policies) and systems (e.g., infrastructure, ecosystems) (Birchall & Bonnett, 2021). Improving the dialogue among planners, designers, and emergency managers can help understanding, in a multi-hazard perspective, the implications of spatial, physical, and functional features of buildings, public spaces and critical infrastructure on the effective management of response operations during disaster risk events (Poljanšek et al., 2021).

CASE STUDY 4 THE SUPERBLOCK PROGRAMME IN BARCELONA[13]

Ione Avila-Palencia and Jole Lutzu

In the 1850s, urban planner Ildelfons Cerdà proposed a plan for a large, grid-like district outside the old walls of Barcelona, called the Eixample ("expansion" in Catalan). Aiming to improve public health, the plan considered the human needs for natural light and ventilation, open space and greenery, and a transport network accommodating pedestrians, carriages, and tram lines. Unfortunately, Cerdà's plan was not realized until the mid–1990s, when Barcelona's first superblock[14] was developed. Since then, Barcelona has built six new superblocks with a plan to have twenty-one by 2030 (Marco,

---

[13] See extended version of case study at https://uccrn.ei.columbia.edu/case-studies.

[14] A superblock is an urban area that encompasses multiple city blocks, where vehicle traffic is restricted to the perimeter, and the interior is designed to prioritize pedestrians with limited car access.

CASE STUDY 4 (cont.)

2024). As Barcelona's air and noise pollutants have reached the highest levels in Europe, superblocks offer an alternative to traditional urban planning.

The superblock model emphasizes the reduction of traffic based on limiting the space dedicated to cars and promoting a transport mode shift offering better public transport service and wider infrastructure for active modes of transport (walking, cycling), as well as an increase in green spaces. Decreasing cars and expanding green spaces reduce GHG emissions and creates a more climate-resilient city.

The superblock is an intervention aiming to reclaim space for people, reduce motorized transport, promote sustainable mobility and active lifestyles, provide urban greening, and mitigate climate change effects. The original idea was to take nine city blocks and close them internally to through traffic. Due to the traffic congestion poorly redistributed in the roads outside the block, the model was recently optimized using green "axes" instead of blocks (City of Barcelona, 2020). Green axes are tree-lined corridors, often with rain gardens, that provide shade, increase air quality, and improve flood drainage systems.

Figure 10 Superblocks in Horta and Sant Antoni (Source: Ione Avila-Palencia & Jole Lutzu, 2023).

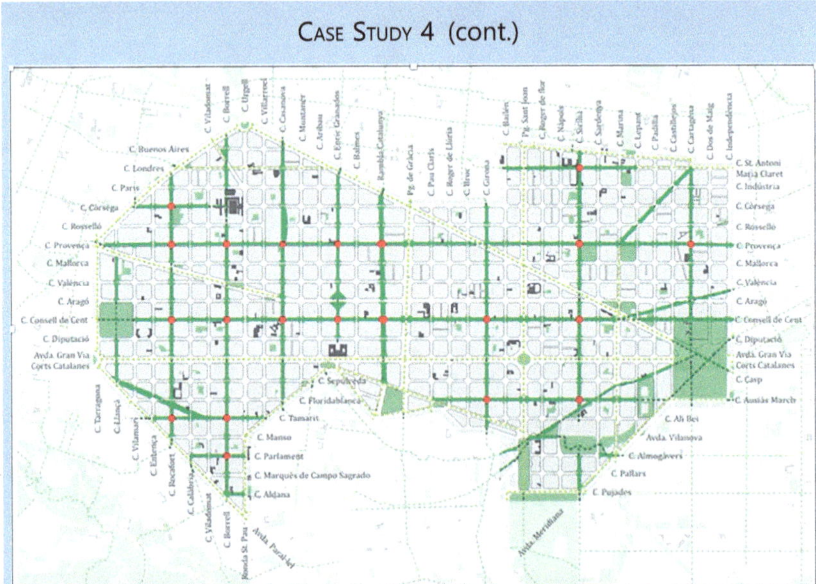

Figure 11 Barcelona Superblocks.
(Source: Zigurat Institute of Technology, 2021).

Between 2013 and 2018, Superblocks were implemented in the areas of Poblenou, Horta, and Sant Antoni. After a controversial lack of participation processes in Poblenou, multi-stakeholder decision-making processes have occurred in working groups that are steering the design of the Superblocks in different neighborhoods. A study published in 2020 suggested that the full implementation of the Superblocks could prevent 667 premature deaths annually in Barcelona because of reducing air pollution, noise, heat, and increasing physical activity levels and access to green space (Mueller et al., 2020).

Superblocks are now being implemented in several cities in the world, including Vienna, Los Angeles, Bogota, and Rotterdam (IAA Mobility, 2024). However, they can still encounter barriers and result in negative outcomes that could limit their development and effectiveness. Future research should evaluate potential consequences such as risk and effects of gentrification and the current housing crisis.

5.3 Approaches

While mitigation and adaptation actions reduce risks associated with climate change, they have mostly operated as separate paradigms. Historically, climate action planning focused primarily on mitigation, with a main goal of preventing climate change tipping points (Watkiss et al., 2015). In response to such priorities, mitigation efforts in planning, urban design, and architecture mainly resulted in strategies to increase energy efficiency of buildings and the share of energy production from renewable sources, while strengthening public transportation and promoting sustainable mobility. Adaptation planning, long neglected in cities, is now increasing rapidly. Cities, out of necessity, are now addressing growing impacts from climate hazards such as heat waves, intense precipitation events, coastal and riverine flooding, hurricanes and typhoons, droughts, and wildfires.

While mitigation targets focus on a single global metric, that is, GHG emissions (Watkiss et al., 2015), adaptation and resilience are primarily concerned with impacts that are local or regional and have a range of non-objective metrics to assess impacts and adaptation responses (Christiansen et al., 2018). (See Additional Resources, Table 1, for characteristics of adaptation and mitigation measurements.)

Mitigating climate change to achieve the 1.5°C global goal of limiting warming requires a drastic reduction in GHG emissions in urban areas from buildings, transportation, industry, and all material streams entering and leaving the city. These are Scope 3 emissions in the GHG Protocol (De Abreu et al., 2022). Carbon reduction targets and carbon pricing can drive retrofit programs and have led to increased building efficiency standards in provincial and national building codes (e.g., the NYStretch Energy Code 2020 in US, or the Level(s) system in Europe) (NYSERDA, 2019; European Commission, 2021a).

Municipal governments often set the pace through ambitious retrofit programs of their own assets, while the private sector and homeowners are often reluctant to pay upfront investment costs with long payback times. Innovative financing concepts and bundling of assets to achieve economies of scale could be solutions to such barriers (e.g., Energy Performance Contracts (EPCs) and Renewable Energy Communities (RECs)).[15]

Some mitigation strategies also support adaptation to climate change, but the achievement of climate benefits strongly depends on the quality of urban design. For example, urban greening initiatives can contribute to carbon sequestration while reducing flood and heat impacts, but require careful design with the support of ecologists and hydrologists to ensure that targeted ecosystem services are actually realized over time. In October 2020, the city of Izmir, Turkey, opened its Mavisehir Peynircioglu Stream Ecological Corridor, a 41,000m^2 greenbelt (European Commission, 2021b). The project aims to

---

[15] See ARC3.3 Element on *Financing Urban Transitions to Climate Neutrality and Increased Resilience in Cities, www.cambridge.org/core/publications/elements/elements-in-climate-change-and-cities*

sequester carbon, increase biodiversity, reduce UHI, and mitigate flood risk (European Commission, 2021). Implemented as part of an EU-funded project (Urban GreenUP), the corridor was a massive undertaking that required the expertise of a wide range of specialists. To strengthen the outcome of this initiative and lock-in greening without requiring major funding, Izmir constructed several "parklets" throughout the city. These small green areas take up less space than the ecological corridor, offer more cost-effective urban greening solutions, and improve pedestrian flow by repurposing street parking spaces (European Commission, 2021).

With the growing number of climate hazard events, city governments are now allocating resources for adaptation initiatives. Depending upon the context and hazard in question, there are opportunities to align adaptation efforts with decarbonization objectives. However, some crucial adaptation strategies may increase GHG emissions. In Singapore, 99 percent of private condominiums are air conditioned, with many crediting the city-state's economic rise to accessible air conditioning. However, while air conditioning can be viewed as a legitimate adaptation strategy amidst rising temperatures, it increases GHG emissions and anthropogenic heat, thus intensifying UHI and creating a vicious cycle (Yuan et al., 2022; Chen, 2023).

Urban and regional planning is a key strategic tool that can facilitate implementing an integrated approach to both mitigation and adaptation actions, for example, the siting and design of renewable energy solutions (Norman et al., 2021). However, an analysis in Europe indicates that planning for climate change mitigation does not always precede adaptation, with about 66 percent, 26 percent, and only 17 percent of cities, respectively, having mitigation, adaptation, or joint plans in place (Reckien et al., 2018b). (See Additional Resources for an analysis of synergies, trade-offs, and conflicts of adaptation and mitigation approaches.)

5.4 Urban Scales

Integrated mitigation and adaptation in cities can assume many forms across spatial scales, urban systems, and physical networks; a wide range of strategies adopted in service of urban sustainability already advances this objective (Raven et al., 2018). However, recent research indicates that implementation of adaptation planning is lagging behind mitigation action in urban areas, and that overall, adaptation planning is often occurring in a sector-specific way with limited consultation (IPCC, 2022). While there is sparse published research analyzing methods for integrating mitigation and adaptation, they are increasing in practice.

Early academic attention focused primarily on governance frameworks for overcoming the adaptation/mitigation dichotomy rather than identifying and investigating adaptation-mitigation interactions in detail (Tol, 2005). Analysis of adaptation-mitigation interactions was aimed at broadly determining the right policy mix of adaptation and mitigation actions (McKibbin & Wilcoxen, 2004), typically at larger scales (i.e., global and national scales) (Watkiss et al., 2015). As research has evolved, some integrated adaptation–mitigation frameworks have been developed for municipal scales (Walsh et al., 2011; Solecki et al., 2015), community and infrastructure scales (C40 Cities Climate Leadership Group, 2024; Green Business Certification Inc., 2018), and the building scale (Judah, 2020). More broadly, they have been integrated into sustainable development planning (IPCC, 2022).

Sections 5.4.1 to 5.4.4 outline approaches and initiatives to integrate adaptation and mitigation at four scales: metropolitan region, city, (largely focusing on urban planning mechanisms), neighborhood (urban design mechanisms), and building (architecture/design).

5.4.1 Metropolitan Region

The metropolitan scale is increasingly recognized as pivotal for effective climate action, offering a framework for coordinated efforts across municipalities that share infrastructure, resources, and climate vulnerabilities. *Chapter 6: Cities, Settlements and Key Infrastructure* in the most recent IPCC report (AR6) recognizes the complexities and interrelationships between rural and urban settlements and calls for adaptation actions that are not constrained by a city's administrative boundary, rather, recognizing the networks and flows between urban, suburban, and rural areas (Dodman et al., 2022). Metropolitan regions include diversified types of territories, encompassing urban, suburban, and rural communities, with interconnected critical systems such as transport, water, food, waste, and energy, which are primary drivers of carbon emissions.

From a planning and urban design perspective, the metropolitan level provides a unique opportunity for enhancing local mitigation potential, by acting on the decarbonization of such systems from a circular and decentralized perspective, and on adaptation, by coordinating responses to extreme events across multiple municipalities and regenerating ecosystems to exploit their regulating and provisioning services.[16] The New Urban Agenda emphasizes how metropolitan regions can leverage their unique position to foster synergies between local governments, the private sector, and civil society, benefiting from regional planning bodies (UN-Habitat, 2017). However, comprehensive

---

[16] See ARC3.3 Element, *Circular Economies for Cities*, https://www.cambridge.org/core/publica tions/elements/elements-in-climate-change-and-cities

planning at the metropolitan scale is often constrained by fragmented governance structures and differing local priorities (Kern & Alber, 2009). Regions with dedicated metropolitan agencies, such as Greater London Authority, show enhanced capacity for implementing climate policies that transcend municipal borders, enabling a broader reach and impact (Bulkeley & Betsill, 2013).

Regional green and blue infrastructure networks can serve as both carbon sinks and resiliency measures, because they absorb and sequester carbon dioxide, reduce the UHI, and help manage inland and coastal flooding. Such networks of existing and restored sea basins, rivers, lakes, wetlands, canals, and reservoirs (blue) connected with agricultural areas, parks, forests, redeveloped wastelands, and degraded areas (green) in strategic locations across urban and peri-urban areas can provide mutual benefits to local communities and ecosystems supporting quality of life and public spaces in cities and creating stepping-stones habitats enhancing regional connectivity for biodiversity (Lynch, 2018).

Metropolitan networks and inter-municipal organizations can facilitate collaborative climate planning, yet these networks often face challenges in ensuring consistent policy implementation. For example, the C40 Cities Climate Leadership Group has fostered collaborative climate strategies across metropolitan regions globally, yet disparities in resources and governance models can create barriers to comprehensive action (Acuto, 2016). While initiatives such as these demonstrate the potential of coordinated metropolitan efforts, they also highlight the challenges that metropolitan regions face in aligning diverse local and regional objectives.

A significant gap remains between policy creation and implementation at the metropolitan scale. Although metropolitan regions may adopt climate policies, local implementation often falls short due to competing priorities and resource limitations (Aylett, 2015; Measham et al., 2011). This highlights the need for stronger mechanisms to translate metropolitan climate commitments into concrete plans and projects able to deliver tangible actions on the ground. The underlying challenge of adopting effective climate-resilient planning and urban design approaches at the metropolitan scale is the difficulty of meeting the varied needs of local communities while addressing region-wide complexities, supporting strategies that allow flexibility and tailored responses across jurisdictions.

5.4.2 City

The way in which a city is defined varies significantly across the world, based on population size, physical size, governance, and location. Cities are defined as having a government that has the authority to legislate and implement climate change plans.[17]

Higher urban density, compact urban form, and mixed land use can facilitate balance between commerce and housing and encourage the local supply of daily needs. The reduction of travel distance also reduces residents' unnecessary motorized travel and encourages walking, cycling, and use of public transportation (Wang et al., 2018). Enhanced public transportation has the effect of reducing both GHG emissions from single-occupant vehicle use and waste heat emissions that contribute to the UHI effect. It is also important to promote use of non-fossil fuel vehicles.

Urban green and blue spaces can improve water quality, create wind corridors, reduce impacts of flooding, and alleviate heat waves and droughts. These reduce GHG emissions, enhance carbon sequestration in vegetation, and cool cities through evapotranspiration and shading (Demuzere et al., 2014). Decentralized energy systems with renewable energy facilities can reduce fossil-fuel energy consumption and increase energy efficiency. Improving the recycling of solid waste and wastewater can also effectively decrease emissions.

There is great potential to integrate adaptation and mitigation goals at the city scale. At present, research indicates integration at the city scale is not widely occurring (IPCC, 2022). For example, in a study of 885 city climate change plans in Europe, Reckien and colleagues (2018b) found that only 17 percent have a plan that integrates both adaptation and mitigation, with 66 percent having a mitigation plan only and 26 percent an adaptation plan only. Grafakos and colleagues (2018) analyzed the climate change adaptation plans (CCAPs) of nine cities across the world and found that six combined adaptation and mitigation: Bangkok, Chicago, Mexico, Montevideo, Seoul, and Wellington. Additionally, Reckien and colleagues (2018b) found that the format of climate plans and their impetus was influenced by numerous factors including size of a city, their international networks, and applicable national legislation.

Many studies focus on analyzing standalone climate plans (including adaptation or mitigation plans), rather than encompassing a suite of city-level policies holistically. Such analysis was undertaken by Hürlimann and colleagues (2021b) who looked at citywide planning documents (across multiple policies and sectors). The analysis showed that integration of adaptation and mitigation goals were not occurring effectively in Melbourne (Hürlimann et al., 2021b).

While these policy evaluation studies are important, a limitation is their lack of analysis of actual actions that implement those plans. There are important

---

[17] See IPCC Special Report on Climate Change and Cities.

Table 3 Integration of climate change adaptation and mitigation in cities

City	Population	Policies	Approach	References
Melbourne	179,000	- Climate and Biodiversity Emergency Declaration - Emissions Reduction Plan - Climate Change Adaptation Plan - Ten Climate Change Action Priorities - Actions embedded in City actions	Climate change adaptation plan and emissions reduction plan are linked, with multiple synergies.	City of Melbourne (2017, 2019, 2021)
Vancouver	697,000	- Climate Emergency Action - Climate Change Adaptation Plan - Rain City Strategy - Zero Emissions Building Strategy - Zoning and building regulations	Long-term policy development, planning, and action	City of Vancouver (2016, 2019, 2025)
Hong Kong	7,482,000	- Climate Action 2050 - Clean Air for Hong Kong 2035 - Green Tech Fund	Multiple policies, guides, and initiatives	City of Hong Kong (2020, 2021)

Notes: (Population statistics from UN Department of Economic and Social Affairs, 2020)

gaps between climate change policies and their implementation. Table 3 shows examples of how cities are integrating adaptation and mitigation across multiple policies and sectors. Crucially, it is important that the selection of a city's approach to integration of adaptation and mitigation actions considers the local community's needs and characteristics, thus enabling a diversity of approaches (Solecki et al., 2015; IPCC, 2022).

5.4.3 Neighborhood

The neighborhood scales provides an important opportunity for citizens, practitioners, and environment professionals to interact to further integrate adaptation and mitigation. Much of the focus on integrating mitigation and adaptation has been at the national and city scales in relation to renewable energy and electrifying transport networks, both between and within cities. The more fine-grained spatial scale of a neighborhood, a precinct, or even a block enables local communities to be actively engaged in the process of decision-making, design, and implementation (Reid & Huq, 2014; Schipper et al., 2014). Community-based solar farms (Boulder), integrated green infrastructure (Rio de Janeiro), landscaping for city blocks (Barcelona), district energy systems (e.g., in Copenhagen and Singapore), and active travel programs for cycling and walking (Addis Ababa) are all strategies that can reduce GHG emissions and, with the inputs of climate science and research, be designed to better adapt to climate change impacts.

In high-density urban districts, Active Integrated Mitigation and Adaptation (AIMA) infrastructure can be inherently more cost-effective to "upgrade" than individual building systems and can therefore better "climate-proof" the built environment against changing conditions (Raven et al., 2018). AIMA deploys advanced building systems and district infrastructure such as integrated energy management, recycling, and on-site wastewater treatment that enables climate resilient development. The neighborhood scale provides the critical link between the city scale and the building envelope. It also requires cooperation and collaboration among multiple owners and interests to co-design and implement change. Recent examples are Singapore's Tengah Town, combining a district cooling system with smart building features, nature-based solutions and transit-oriented development, or the Western Sydney suburbs greening, that recently banned black roofs on houses. Both examples integrate adaptation and mitigation and were co-designed with leading researchers and local communities.

Compact neighborhood development with efficient density balances urban demand with resources, increases access to public services, and prevents urban sprawl, which can result in increased GHG emissions. Mixed-use areas encourage integrated, inclusive urban systems that are needed to effectively implement mitigation and adaptation strategies, such as increased walkability, the development of green spaces, and improved stormwater management through rain gardens and bioswales. Compact development can also lead to effective sizing of blocks and spaces, non-motorized transport, and accessibility to mass public transit. Efforts to create appropriate urban green spaces with water-sensitive and climate-adaptive design can contribute to energy use efficiency and promote thermal comfort (Demuzere et al., 2014).

The construction of smart grid and renewable energy systems at the neighborhood scale can improve the efficiency and safety of power generation, reduce losses due to transmission and distribution, improve the grid-connected power generation capacity of renewable energy, and improve user management, all of which largely reduce GHG emissions. The "recycle and reuse" system of wastewater and solid waste management can also be closely integrated into residential lifestyles.

Appropriate block form and pavement materials can reduce heat gain and contain excess water. Streets and paths can be designed to accommodate walking, biking, and public transport to help reduce emissions. Green spaces with appropriate plant types can serve as carbon sinks (Zhao et al., 2023). Open spaces can be designed to absorb and retain excess water during flooding. Sustainable devices such as LED lighting can contribute to emission reduction targets. Wastewater recycling and reuse equipment as well as solid waste sorting and collection facilities are to be designed to ensure for the convenience and adoption by users.

5.4.4 Building

A limited number of integrated adaptation-mitigation frameworks currently exist for buildings (see Table 4). C40 Cities developed the Adaptation and Mitigation Interaction Assessment (AMIA) Excel-based tool, where users can select from a database of adaptation and mitigation strategies at multiple scales, including the building scale, to identify potential synergies and trade-offs (C40, 2018).

The LEED® (Leadership in Energy and Environmental Design) voluntary green building certification system, one of the most recognized of such systems worldwide, is for the first time introducing, in its "version 5," an assessment of climate resilient development components, strengthening aspects related to emissions reduction and explicitly including adaptation, spatial justice,[18] habitats, and biodiversity. (See Additional Resources, Table 2, for green building principles in LEED v5 linked to climate resilient development.)

In the US, the FEMA's (Federal Emergency Management Agency) report, Natural Hazards and Sustainability for Residential Buildings (Gromala et al., 2010), considers interactions between sustainable building strategies and adaptation objectives. The guideline looks at the impact of natural hazard risks on specific sustainable building strategies, though impacts of resilience strategies on sustainability and climate mitigation objectives are not investigated. The document also notes how strategies could be synergistic for one hazard but create a conflict for another. Using life cycle assessment (LCA), the document

---

[18] Spatial justice, refers to the fair distribution of urban resources and opportunities, ensuring that people of all socio-economic backgrounds can access and benefit from the urban environment.

Table 4 Integrated adaptation-mitigation frameworks for buildings

Framework	Description
Adaptation and Mitigation Interaction Assessment (AMIA)	Users select from a database of adaptation and mitigation strategies at multiple scales, to identify potential synergies and trade-offs.
LEED v5®	Points-based green building design rating system encompassing carbon neutrality, climate adaptation, social and environmental quality.
FEMA Natural Hazards and Sustainability for Residential Buildings	Guidelines focus on impact of natural hazard risks on specific sustainable building strategies; explains how strategies could be synergistic for one hazard but create a conflict for another.
Resilient Adaptation of Sustainable Buildings	Uses regenerative framework to delve into specifics of meeting an adaptation goal for a predetermined hazard scenario while prioritizing sustainable strategies.
Integrated Building Adaptation and Mitigation Assessment (IBAMA)	Twelve-part process-based framework to assist building developers, owners, and design teams in increasing project's resilience and adaptation to climate hazards while optimizing climate mitigation and sustainability goals and strategies

addresses embodied GHGs and introduces the concept of post-disaster sustainability benefits.

Resilient Adaptation of Sustainable Buildings posits that using a regenerative framework rather than an efficiency-based approach (i.e., focusing on ecosystem services restoration rather than on reduction of energy consumption and GHG emissions only) will result in buildings that are both sustainable and resilient. It serves as a valuable design approach for delving into the specifics of meeting an adaptation goal for a predetermined hazard scenario while also prioritizing sustainable strategies. However, it does not specifically address trade-offs and excludes embodied carbon factors (Graves, 2020).

Passive buildings, which relates to the structure of the building itself, including orientation, window placement, skylight installation, insulation and building materials, and specific elements such as windows and window shades, are another form of integrated mitigation and adaptation (Anand et al., 2023). These approaches can maximize natural light and wind to decrease heat gain, improve ventilation, and minimize energy load (International Passive House Association, 2018). Green roofs and walls can cool entire buildings, reducing the need for indoor air conditioning because green facades and surfaces help cool them via evapotranspiration. Building-integrated photovoltaic (BIPV) can serve as both the outer layer of a structure and generate electricity for on-site use or export to the grid. Wastewater reuse systems and solid waste sorting systems can also be installed in buildings.

Originally developed for multi-unit residential buildings, the Integrated Building Adaptation and Mitigation Assessment (IBAMA) is a twelve-part process-based framework to assist building developers, owners, and design teams to increase a project's resilience and adaptation to climate hazards while optimizing mitigation and sustainability goals/strategies (see Figure 12). The Integrated Building Adaptation and Mitigation Assessment functions as a flexible decision-making tool rather than a checklist or series of requirements. This enables teams to respond to the unique context, vulnerabilities, and circumstances of each project. The framework provides twenty-one quantifiable evaluation criteria to determine

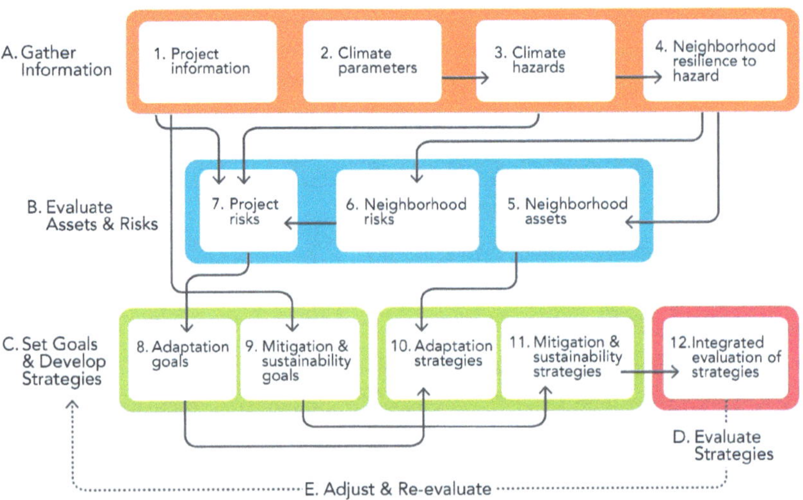

Figure 12 IBAMA framework guides a project's approach to climate adaptation and resilience and includes consideration of the neighborhood scale (Judah, 2020).

synergies, conflicts, and trade-offs among all proposed adaptation and mitigation strategies, as well as alignment of the proposed strategies with other project priorities such as cost, constructability, and ease of operations (Judah, 2020). (See Additional Resources, Figure 7, for IBAMA evaluation criteria.)

6 Embedding Environmental and Climate Justice in Planning and Design

Environmental justice is gaining a central role in planning and urban design because of its urgency and complexity (Porter et al., 2020). There is clear evidence that further delays in emissions reductions will further limit effective adaptation, exacerbating losses and damages, and greatly worsening not only climate change impacts, but also climate injustices (Porter et al., 2020; IPCC, 2022). Inclusion, equity, and justice dimensions are increasingly considered a key to shaping climate-resilient development pathways (IPCC, 2022).[19]

An increasing volume of literature is dedicated to understanding the implications of climate justice at the urban scale. However, there is limited awareness about how to effectively embed environmental and climate justice in climate-resilient urban planning and design in practice, as justice concerns are only recently emerging in different cities worldwide (Bulkeley et al., 2014). Most of the studies conducted in this field remain on a theoretical level, with empirical studies mostly focusing on justice integration at the procedural level of planning and on the impact of cities' climate plans on inequality (Fitzgibbons & Mitchell, 2019; Klinsky & Mavrogianni, 2020). These studies contribute to framing how justice and equity are currently considered in city planning and district redevelopment projects.

Findings show that only a few cities have integrated justice and equity effectively in climate action (Hess et al., 2021). Significant gaps are found in how planning and urban design can help overcome inequitable socio-spatial outcomes of both climate change impacts and urban transformation. The most affected are the least responsible for global warming and have fewer resources to cope with and recover from climate impacts, while simultaneously climate change multiplies existing urban injustices (Reckien et al., 2018a; Foster et al., 2019; Long & Rice, 2019).

Thus, solutions need to embed transformative pathways capable of integrating processes of empowering equity-driven urban land and housing policies and collaborative and inclusive planning to reduce urban dynamics of inequality by including vulnerable groups (Shi et al., 2016; Rosenzweig et al., 2018; Shi, 2021).

Equity in urban planning and design practices cannot be achieved without fundamental change at both political and practitioner levels (Klinsky &

---

[19] See ARC3.3 Element, *Justice for Resilient Development in Climate-Stressed Cities*, www.cambridge .org/core/publications/elements/elements-in-climate-change-and-cities

Mavrogianni, 2020). Similarly, moving from an incremental mitigation approach to concurrent decarbonization and adaptation will require a system-wide shift in urban development processes (Tozer et al., 2018). Urban planning and architecture practitioners as well as policymakers who aim to achieve a just climate transition could therefore pursue systemic change in urban planning and development processes, understanding how urban form and governance influence the distribution of climate risks and vulnerabilities, and how to advance participatory methods (Hughes & Hoffman, 2020).

This section identifies emerging practices in pursuit of climate justice into urban development processes and possible pathways for just climate-resilient design dealing with elements of recognition, rights, responsibilities, distribution, and procedures (Archer et al., 2014; Chu et al., 2016).

6.1 Intersection of Climate Policy, Urban Planning and Design, and Justice

Integrating climate justice into climate policymaking and planning can help avoid the perpetuation of social inequities in cities. As seen in Table 5, pioneer cities have recognized this opportunity and have incorporated climate justice into their climate action plans creating intersectional climate and land use policy aimed explicitly at addressing inequality and vulnerability (Table 5) (Diezmartínez & Short Gianotti, 2022).

Conversely, policy focused too narrowly on climate benefits has failed to address social inequalities, and in some cases exacerbated vulnerabilities in historically marginalized populations (Bouyé et al., 2020).[20] Climate policy and planning decisions that do not explicitly prioritize procedural and distributional climate justice have been criticized as technocratic and exclusionary (Ajibade, 2019). This has alienated local stakeholders and slowed climate progress (Cohen, 2018).

Examples of these outcomes can be found in urban densification policies that are driven by environmental benefits alone. Resulting projects have exacerbated spatial inequalities, such as unequal access to green space, a phenomenon which can be observed in cities in both the Global North and South (Rigolon et al., 2018; Schüle et al., 2019). This phenomenon can be seen in some cities with left-leaning politics and progressive sustainability policies. In Oslo, while compact city planning policy efforts were often located in lower income and immigrant districts with low provision of blue-green space (Venter et al., 2023).

---

[20] See ARC3.3 Element, *Justice for Resilient Development in Climate-Stressed Cities*, www.cambridge
     .org/core/publications/elements/elements-in-climate-change-and-cities

Table 5 Climate plans and actions of cities that are developing measures to achieve climate justice outcomes

City Climate Plan	Focus
Los Angeles Green New Deal Plan	Reduction of emissions and inequality through low-carbon housing, union-led green jobs, and resilience in informal settlements
Cape Town's ceiling retrofit project	Roof insulation in low-income areas improves health, lowers emissions, and reduces financial stress
New York City's Cool Neighborhoods program	Heat mitigation in high-temperature neighborhoods through reflective surfaces and green areas; implemented with community groups
Barcelona's Climate Plan	Focuses on climate justice through stronger care centers and extreme-heat shelters for vulnerable groups
Buenos Aires' Hydraulic Master Plan	Flood mitigation around 8 water bodies, prioritizing low-income Lake Soldati area
Portland and Multnomah County's 2015 Climate Action Plan (CAP)	Guided by community reps from low-income areas; advances procedural, distributional, and structural equity through local investment, job creation, and climate equity metrics
Sunset Park in Brooklyn, New York	After Superstorm Sandy, residents worked with UPROSE to create a neighborhood climate adaptation and environmental justice plan

Climate-oriented development projects have often focused too narrowly on climate benefits, failing to address the needs of local stakeholders and thus garnering resistance. A São Paulo urban densification project, Nova Luz, exemplifies this. The project was envisioned in response to the city's newly enacted emissions reduction law in 2011. Despite the promise of affordable housing, the project was met with resistance by those occupying the site, which grew to include resistance by the broader housing movements in the city. These groups argued that affordable housing had previously failed to materialize in urban development projects that did not identify it as the main objective. The subsequent administration proposed a densification plan designed explicitly to tackle inequality, which received broad community support from the same housing movements. This plan reduced emissions even further than the previous plan, though this was not a part of the messaging (Cohen, 2018).

An increasing body of literature has linked urban greening policy to displacement, primarily as the result of the increased real estate value phenomenon often termed "green gentrification" or "climate gentrification." (Oscilowicz et al., 2025). In collaboration with local policymakers, urban planners can play a key role in identifying anti-displacement strategies, such as preserving existing affordable housing and business ownership and demanding rent control through local value capture mechanisms (Rigolon et al., 2018; Oscilowicz et al., 2022).

Climate policy that distributes funding is most effective when carefully designed to meet distributive, procedural, restorative, and recognitional justice objectives (See Case Study 5).[21] Energy incentives are particularly susceptible to

---

CASE STUDY 5 CITY-LEVEL CLIMATE ACTION PLANNING IN KANO, NIGERIA

Enjoli Hall, Aliyu Salisu Barau, and Darien Alexander Williams

Kano, Nigeria, is the southernmost trans-Sahara trading city and has roughly four million residents who are predominantly Hausa and Fulani language-speaking people, with strong Indigenous and Islamic cultural and spiritual heterogeneity. Rising temperatures, prolonged droughts, and severe flooding events pose threats to the city's economy, human settlements, and heritage sites, including its ancient city walls and the Great Mosque. Unlike Lagos, Kano does not have a comprehensive state-led strategic plan for climate mitigation and adaptation, and the Urban Planning Authority's current stance is that climate strategies are not a priority. For these reasons, Kano is an example of how local adaptation efforts can reaffirm precolonial Indigenous knowledge systems and meet environmental justice objectives.

Community-based actors in Kano are responding to observed climate change impacts in the absence of formal plans. "Barefoot planners," that is, local actors who have no formal policy authority, draw their power from ordinary, incremental, and persistent practices of adaptation, such as tree planting, community-scale water management, and mutual aid networks to

---

[21] See ARC3.3 Element, *Justice for Resilient Development in Climate-Stressed Cities, www.cambridge .org/core/publications/elements/elements-in-climate-change-and-cities*

repair damaged infrastructure. Barefoot planners often act as first responders to the most devastating impacts of climate change in Kano, forming grassroots groups that pool resources and labor to repair damaged infrastructure in informal settlements in the wake of disastrous floods (Barau, 2020).

For example, the *Millennials and Resilience: City, Innovation and Transformation of Youths Laboratory (MR CITY Lab)* is an initiative aimed at engaging young people to restore urban biodiversity in Kano through planting of local tree species. The team – composed of university students, local youth, scientists and researchers, practitioners, and community leaders – has planted more than 400 Indigenous trees in communities across the city and facilitated communal dialogue about land and flora, built skills, and heightened gender equity. The project has served as a point of decolonial capacity-building to address the erasure brought by generations of colonial rule, as knowledge of local flora is shared. It highlights the important role of informal knowledge, actors, and practices in local climate action planning. It remains to be seen, however, whether the micro-practices of nongovernmental actors encourage coordinated, citywide climate action in Kano.

Figure 13 Left: University students participating in tree restoration activities with the MR CITY Lab.
(Source: MR CITY Lab)
Right: Typical mud-built homes in Kano, Nigeria, seen from Dala Hill (Source: Tadej Znidarcic, 2015).

an overly narrow focus on technical criteria and missed opportunities to ensure fair and just allocation of funds. Though minority and low-income populations disproportionately experience energy poverty, energy incentives and associated projects have predominantly served high-income populations. Examples include access to electric-vehicle-only carpool lanes and charging infrastructure, public bike shares, spatial inequities in renewable energy project funding and siting, and uneven distribution of building retrofit incentives (Yenneti et al., 2016; Babagoli et al., 2019; Sovacool et al., 2019; Hsu et al., 2021).

In these examples, a utilitarian approach to subsidy allocation is economically inefficient and disempowers energy-poor communities. Effective regulatory frameworks that include low-income groups with a focus on energy retrofitting in housing have been developed in the UK with its Energy Act and the Minimum Energy Efficiency Standard. This policy gives dwellers a voice in decision-making related to energy retrofitting (Klinsky & Mavrgianni, 2020).

6.2 Approaches for Just Climate Resilient Planning and Design

In many cases, climate policy is developed at the city, regional, national, or even international level, but is ultimately interpreted and carried out at the neighborhood scale by urban planners and urban designers. Appropriate strategies for achieving just climate objectives are therefore best rooted in the local context. Climate priorities that are coupled with local needs by contextualizing climate-aligned urban design and planning measures to places and people are likely to be more effective.

Fundamental tools for the translation of this aim into the urban form are equitable spatial distribution of green and blue infrastructures, adaptive land/building uses, and inclusive and collaborative planning and urban design models. Developing novel planning and urban design methods that link urban climate science with community engagement in collaborative mapping, design, and evaluation of solutions can help to achieve procedural, recognitional, and distributive justice while legitimizing and recognizing the needs of those most exposed and vulnerable to climate risks (Mohtat & Kirfan, 2021).

For example, in 2019, UCCRN conducted a district-level UDCW program in Isipingo, a town outside Durban, South Africa. To facilitate participating City Teams to incorporate climate change initiatives into the Isipingo Rehabilitation Programme (Figure 14). The purpose of the UDCWs was to develop implementation actions that considered governmental, developmental, socioeconomic, and ecological conditions and build capacity across multiple stakeholder groups to respond to climate change. The UDCW enabled

Figure 14 Community engagement session: Finding synergies between community and climate priorities for the Durban Isipingo central business district (CBD)

(Source: Enza Tersigni, 2019).

city officials from eThekwini Municipality to integrate and scale up mitigation and adaptation principles by reducing energy consumption, strengthening resilience, and enhancing human well-being. The group worked synergistically with construction and landscape configurations to create equitable, interconnected, protective, resilient, and attractive urban areas. (See the additional resources for more details on the Durban UDCW).

Transformation of the built environment is key to leveraging the reduction of existing socio-spatial inequalities and root causes of vulnerability, but this can be achieved only conjunctly with more democratic and participatory governance and spatial decision-making (Wolfram et al., 2019; Hughes & Hoffmann, 2020; Castan Broto, 2021). Of paramount importance is the integration of inclusiveness and climate action through embedding the claims and needs of the populations usually excluded from spatial decisions due to race, gender, income, culture, religion, or other social factors of exclusion. Urban services and the environmental benefits and co-benefits generated by climate-resilient planning and design measures deliver a major opportunity to address inequalities in vulnerable groups and areas.

A critical component of a collaborative and inclusive systematic approach is active and effective stakeholder and community engagement. Climate equity and justice outcomes have been shown to increase through inclusive planning and collaborative design processes (Chu, 2016). Without such engagement, both climate and equity goals may be undermined due to lack of public support for implementation and lack of ownership to encourage sustained equitable low-carbon transition outcomes. Moving from a top-down toward a participatory climate planning and execution process is thus an important priority for policymakers and practitioners alike. McCauley et al. (2019) highlighted that this is particularly important in cities experiencing rapid and informal expansion.

Different cities have employed a range of public engagement methods from consultation to co-design, co-production, and co-dissemination of results within their climate plans (Table 6, Bremer et al., 2019; Satorras et al., 2020).

An approach recurrent in climate justice literature is partnering with community-based organizations and grassroots movements. The organizations and movements provide expertise and leadership within the communities they serve, facilitating implementation of climate resilience projects (Gonzalez et al., 2017; Urban Sustainability Directors Network, 2017; Amorim-Maia et al., 2022). They enable

Table 6 City climate plans with participatory processes for urban climate planning (See Additional Resources for expanded version)

City Climate Plan	Participatory Process Implemented
Detroit Climate Action Collaborative (DCAC) (US)	Prioritized community involvement as key aspect of plan's formulation; incorporated local insights and increased climate change awareness in Detroit; organized range of activities such as focus sessions with stakeholders, local climate conference, targeted consultations with commercial and religious groups, and documentary films; process led to creation of Detroit Climate Ambassadors program, a resident-led effort to educate, prepare, and tackle climate change in neighborhoods
Cleveland Climate Action Plan (US)	Year-long community engagement process that developed 2018 CAP Update; 300 residents participated in 12 neighborhood workshops, city received more than 200 comments during public comment period; Climate Action Advisory Committee was created with more than 90 members include representatives from business, churches, academia,

Table 6 (cont.)

City Climate Plan	Participatory Process Implemented
	NGOs, the city and county, community organizations, and foundations
New York's participatory budgeting process (myPB) (US)	Facilitation process for investment of more than $200 million in 700 community-designed projects
CityAdapt Project, San Salvador, El Salvador, and Xalapa (Mexico)	Funded by GEF and implemented by the United Nations, which developed climate adaptation plans for San Salvador, El Salvador, and Xalapa, Mexico; worked hand-in-hand with communities to identify vulnerability hot spots; conducted participatory workshops to identify possible and necessary actions; conducted workshops to validate portfolio of actions; presented results to decision-makers to ensure support and; created monitoring and evaluation system in collaboration with communities
Sustainable Food Production for a resilient Rosario (Argentina)	Repurposed under-utilized land for urban and peri-urban agriculture to improve food security, provide nutrition to low-income residents, and strengthen resilience to flood and extreme heat
Climate Resilient Cities in Latin America Initiative, Dosquebradas, (Colombia); Santa Ana, (El Salvador); and Santo Tomé (Argentina)	Developed practical way to integrate stakeholders into decision-making process using specific tools; Tools included the QUICKScan methodology and decision-support toolbox developed by Wageningen Environmental Research (WEnR/ Alterra) and the European Environmental Agency (EEA); QUICKScan methodology has been

Table 6 (cont.)

City Climate Plan	Participatory Process Implemented
	implemented to integrate different data, knowledge bases and perspectives, and needs of stakeholders; stakeholder engagement created a Climate Resilient Cities Initiative platform and workshops with stakeholder mapping, interviews, policy recommendations, and capacity development
Barcelona Climate Plan (Spain)	Elaborated through "co-production" process with citizens, which consisted of three phases; 1) Collected proposals from citizens through face-to-face workshops, self-organized sessions, and digital platform Decidim; 2) Validated and prioritized proposals through municipality-organized workshops; 3) Evaluation and acceptance or rejection of proposals with decisions reasonings posted on Decidim platform
Paris "Oasis Schools" (France) and Barcelona "Escuelas refugios climaticos" (Spain)	Retrofitted schools as climate shelters; school emerged as neighborhood-scale climate adaptation and community hub; from pilot projects, initiatives turned in urban policies based on collaborative design processes with local community

design and development of adaptation projects from within communities, promoting jobs, skills creation, and achievement of community-driven adaptation (Gonzalez, et al., 2017; Amorim-Maia et al., 2022).

The place-based and place-making approach involves integrating within projects the relationship of different communities with space, including recognition of vernacular and local knowledge especially of Indigenous communities

(Anguelovski et al., 2018). It also implies a decolonial approach to land distribution and access, including return and redistribution options. Additionally, promotion of cross-identity and vulnerability activism can strengthen government actions that are already established in the community, empowering local organizations to manage changes (Olazabal et al., 2021).

Community-Based Adaptation (CBA) is a complementary framework for just climate-resilient urban planning and urban design that focuses on vulnerable communities through participatory activities at multiple levels (from national to neighborhood scales) in a wide range of spheres (project implementation, policy, education, and research) (Ayers & Forsyth, 2009; Forsyth, 2013; Archer et al., 2014; Kirkby et al., 2018; Rosenzweig et al., 2021). In this framework, the analysis and assessment of vulnerabilities and risk, as well as response measures are developed according to a participatory learning and action model (Kindon et al., 2007).

In CBA, communities are the central focus in both studies of climate risks and vulnerability and in spatial decision-making (Kirby, 2014). Culture can play different roles in community-based adaptation, for example, including local social norms, effecting change from within, entraining adaptation (see Box 3).

Box 3 THE ROLE OF CULTURE IN IDENTIFYING URBAN STRATEGIES FOR CLIMATE ADAPTATION

As urban climate solutions are shared and mainstreamed worldwide, the ability to successfully adapt planning and design to different contexts has become crucial. This is essential not only to ensure the successful implementation of climate action in urban settings but also to account for the diverse communities that exist in local contexts. Culture relates to climate and environmental justice because "all understandings of the environment are politicized and are framed by broader power relations and political-economic structures and processes" (Leck, 2017). In this sense, cultural knowledge that comes from historically oppressed groups can be systematically ignored in adaptation and mitigation strategies, while consequently presenting valuable tools for promoting climate and environmental justice.

In terms of urban planning and design, culture is important because spatial places are relevant to how groups and individuals self-identify (Heimann & Mallick, 2016). Moreover, the identity of social groups can be related to spatial elements from their surroundings such as water sources, landscape elements, or native ecosystems. As climate adaptation can encompass relocating communities or changing conditions concerning land or water sources,

Box 3 (cont.)

culture can inform why certain groups may prefer different adaptation strategies. Since culture can also exist and change at different scales, it is important to consider the different scales of urban planning in climate adaptation strategies (Leck, 2017).

Establishing "best-practices" for adaptation does not usually account for cultural differences. To think of adaptation measures in a global setting, it is necessary to consider cultural, ecological, and institutional contexts (Ensor & Berger, 2009; Heimann & Mallick, 2016). Adaptation strategies should not be imposed from the outside but rather have the complete engagement of communities to drive adaptation action from within culture. From a planning and design perspective, this entails adapting best practices from cities with diverse cultural contexts to local solutions and considering stakeholder and community engagement processes through a cultural lens when adapting these solutions.

This approach promotes a productive role of culture, where shared knowledge from communities defines the opportunities for adaptation, predicts best practices, and builds resilience (Ensor & Berger, 2009). Consequently, culture can provide a series of habits, skills, and styles that can create an "adaptation driven by culture" toolkit (Leck, 2017). Considering adaptation strategies without addressing socio-cultural beliefs can lead to ineffective and resisted policies, and maladaptation.

Practical application of planning tools for CBA have been conducted mainly in developing countries with justice outcomes including inclusive socioeconomic measures, land use modifications, and spatial interventions (Spires at al., 2014; Endo et al., 2017; Kim et al., 2018, Han et al., 2025). Cape Town, South Africa, is one example of a city where CBA is occurring, as local community-based organizations are rearranging homes in informal settlements to allow for flood drainage and service delivery amid increasing rainstorms (Fox et al., 2021).

6.3 Future Research and Practice for Advancing Urban Climate Justice

Urban planners and urban designers can serve as the link between climate policy and environmental justice outcomes, leveraging policy and spatial interventions as tools for increasing wealth and prosperity in marginalized communities. This requires a shift in the standard approach to both climate-oriented and equitable development practices and acknowledging different forms of knowledge (see Box 4). To ensure these outcomes, urban climate justice ought to be elevated to the primary objective and driving purpose, while emissions reductions toward climate policy compliance can be seen as a baseline requirement for all development projects. Business-as-usual planning and urban design approaches will not achieve both climate and equity goals in tandem and should therefore be re-examined. Elements to be better considered in future planning and design practices include rectifying vulnerabilities by tackling underlying economic reinforcers of racial and gender inequalities, as well as adopting place-based and co-produced approaches and promoting community resilience-building and activism (Amorim-Maira et al., 2022).

BOX 4 TRADITIONAL ECOLOGICAL KNOWLEDGE (TEK) IN URBAN PLANNING AND DESIGN

Traditional ecological knowledge (TEK) is understood as the "knowledge of the environment that is derived from experience and traditions particular to a specific group of people" (Leonard et al., 2013, 2). Although TEK is usually associated with ancestral communities, it can be attributed to communities that have historical continuity in a particular environment, with particular resource use (Leonard et al., 2013). In the context of TEK, people's resilience is permeated by local social networks or traditional community structures that constitute the "social capital" for adaptation (Heimann & Mallick, 2016).

When including TEK in urban planning and design, it is possible to gather specific information for implementation associated with the traditions of the communities that will be affected by the various proposed interventions. In this regard, it is feasible to design community engagement based on the different categories of TEK.

For diagnostic and planning purposes, efforts can be focused on environmental knowledge, which relates to seasonal indicators, water availability, and, in general, various observations of climate variability and change (Leonard et al., 2013). For Monitoring, Evaluation, and Learning (MEL), consideration of biodiversity resource usage and environmental management can ensure the long-term sustainability of the interventions. Finally, the worldview should serve as the framework in adaptation solutions are tested for different communities, understanding how their cultural identity interacts with their ethics and values concerning urban climate action (Leonard et al., 2013).

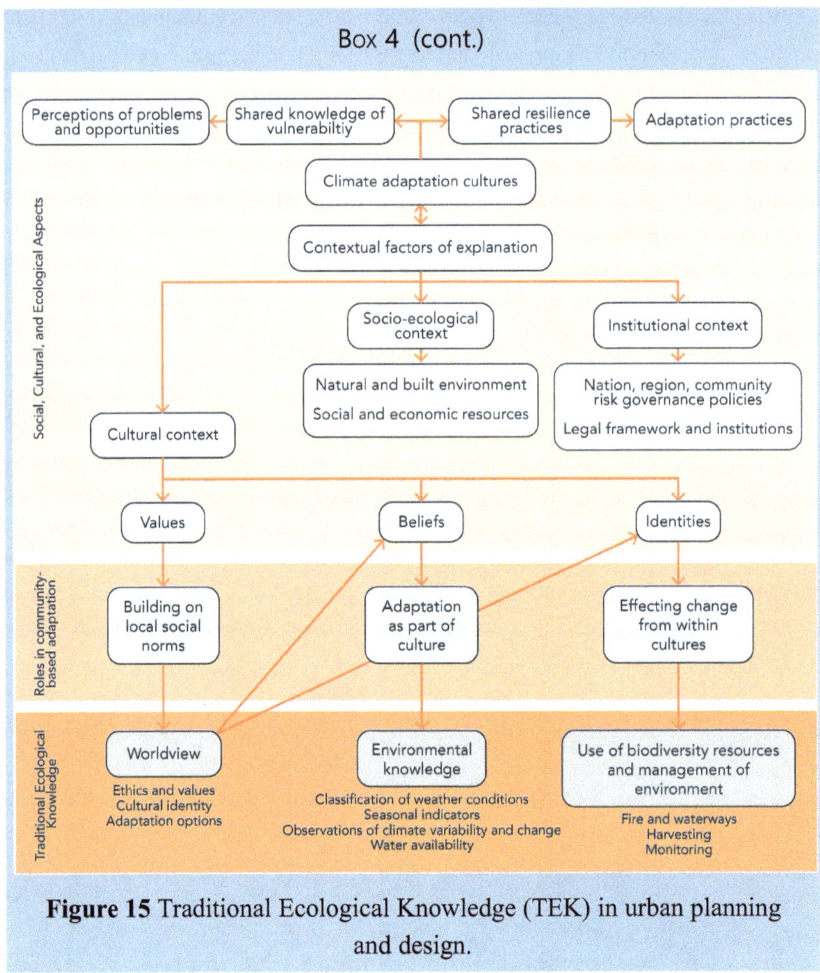

Figure 15 Traditional Ecological Knowledge (TEK) in urban planning and design.

Rectifying vulnerabilities includes diversifying funding sources for climate projects. Studies show that a reliance on private funds has caused investment and real estate firms to speculate on land values, often generating further marginalization and intra-urban displacement (Teicher, 2018; Robin & Castan Broto, 2021). This is exemplified in Chicago, where publicly owned lots in gentrifying areas are often sold to create green spaces (Riglon et al., 2020). Unfortunately, the benefits of these new green spaces are rarely distributed equitably, raising property values and often displacing already marginalized communities (Riglon et al., 2020).

To avoid climate gentrification, unequal land use and zoning policies need to be rectified (Anguelovski et al., 2018b; Conolly & Anguelovski, 2021). An example is the need to ensure affordable housing, while simultaneously ensuring access to green spaces (Anguelovski et al., 2018b). Vienna is a city that seeks to mitigate the risk of green gentrification. The city has a broad dispersion of social housing across neighborhoods to prevent segregation and isolation and has strict rent controls and open-ended contracts that limit price increases (Friesenecker et al., 2024). These regulations have made social housing more equitable and ensure that rents stay consistent despite changing neighborhood demographics. A 2015 plan also introduced an initiative to provide all Vienna residents with green spaces within 250 meters, guaranteeing equitable access to nature (Friesenecker et al., 2024).

Innovating participatory practices in planning and urban design towards inclusive co-production processes requires governance and advocacy through urban policies tailored to the vulnerability of certain groups. Knowledge sharing is needed so that reliable data about climate risks, vulnerabilities, and systemic barriers are available to all.[22] Training to build awareness of inequities and social injustices and to facilitate dialogue among grassroots organizations, NGOs, communities, and policymakers is also required (Klinsky & Mavrogianni, 2020; Mohan & Muraleedharan, 2025).

The need is rapidly emerging for much greater attention to be paid to advancing understanding of equity and justice implications of urban form and of functional, spatial, environmental, and technological outcomes of plans and projects. The lack of empirical studies on the subject has long been coupled with an excessive focus on normative recommendations and critical analyses, which too often fail to expound upon the practical methodologies for propelling climate justice objectives (Hughes & Hoffman, 2020; Mohtat & Kirfan, 2021). More research and exchange between theory and practice are required to assess the results in terms of justice of changes in the size, orientation, geometry and layout patterns of streets, blocks and plots, buildings and their footprints, as well as changes in land use.

Appropriate methodologies belonging to the spatial dimension can be implemented to foster the evaluation of existing urban injustice and the enhancement of equitable interventions in the built environment. These include spatial analysis and on-site measurements with a specific focus on participatory GIS (see Section 8); the latter are used to co-survey and co-own local spatial knowledge of different groups (Korpilo et al., 2022; Rice-Boayue, 2025). This holds particular significance in the Global South, where limited access to spatial data is a gap in the implementation of urban climate justice actions.

---

[22] See ARC3.3 Elements, *Urban Climate Science: Knowledge Base for City Risk Assessments* and *Resilience, and Justice for Resilient Development in Climate-Stressed Cities*, www.cambridge.org/core/publications/elements/elements-in-climate-change-and-cities

Moreover, the use of indicators for the evaluation of justice and equity in cities can support spatial planning and urban design in the elaboration of proposals that are aware of the specific context. Examples of these indicators are the Equity Indicators tool developed by the CUNY Institute for State and Local Governance (ISLG), available for six US cities, and the Environmental Justice Index elaborated by the Centers for Disease Control and Prevention and Agency for Toxic Substances Disease Registry (US Department of Health and Human Services). Toolboxes for Community-based Adaptation (CBA) such as the Cooperative for Assistance and Relief Everywhere (CARE) and the CBA Toolkit are also useful tools to help in delivering justice at a practical level in the built environment (CARE, 2010).

Since urban transformation can actively reduce existing socio-spatial inequalities, the demands of vulnerable groups and individuals, especially of marginalized and discriminated communities, are key to deliver climate-resilient cities. When specific studies on root causes and factors influencing local vulnerabilities are included as baselines in planning and urban design proposals, they can be joined with urban climate analyses to match mitigation and adaptation goals with local priorities. These contextualize climate strategies and measures to people and places. Risks of maladaptation generated by the creation of environmental benefits that favor high-income groups (climate elites) at the expense of the more disempowered, such as green gentrification, can thus be better understood and reduced. Multi-stakeholder, multilevel, and context-based urban processes that embed a justice and equity agenda can help transform how spatial decisions are taken and implemented, how rights are guaranteed, and how responsibilities for climate and community action are undertaken.

7 Capacity Building for Urban Decision-Makers and Practitioners

Achieving global climate goals and disrupting the status quo will depend on the capacity of built environment practitioners working in countries of all stages of development to act on climate change and incorporate it into their professional activities. The Paris Agreement (Article 11) recognizes the importance of capacity building to achieve its goals, with a particular focus on enhancing the capacity of Parties from developing nations, and the role of developed nations in supporting the achievement of this aim (UNFCCC, 2015). During COP21 (Paris, 2015), the UN established the Paris Committee on Capacity-Building (PCCB) to address gaps and needs in developing countries. Further, in 2019 it was decided that the PCCB should also serve the Paris Agreement and develop a work program to advance capacity building (UN, 2023a; UN, 2023b).

7.1 Defining Capacity and Capacity Building

In a report for the World Health Organization (WHO) on capacity building focused on health systems, Milèn (2001) defined capacity across three levels, as "an ability of individuals, organizations, or systems to perform appropriate functions effectively, efficiently and sustainably." This is consistent with the UNDP's (1998) earlier work on capacity building. The definitions of the term "capacity building" and "capacity development" have limited consensus (Klinsky & Sagar, 2022). As detailed by Nautiyal and Klinsky (2022), capacity building for the environment gained prominence in the 1990s through multilateral environmental agreements and continues to have prominence at international forums to date. They identify problems with the use and implementation of the concept. Through an analysis of UNFCCC documents, they identify two dominant narratives of capacity building.

The first narrative was characterized by "tecno-managerial and standardized data-driven goals," which is well supported by the UNFCCC (Nautiyal & Klinsky, 2022). The second narrative, which they found was not well supported, was capacity building that is inclusive and diverse, implemented through holistic and transdisciplinary approaches. Nautiyal and Klinsky (2022) advocate that to increase capacity building for climate change, greater support for the second narrative is needed. In the context of planning and design practices, collaborative processes for knowledge sharing and co-design in multi-stakeholder contexts are key to embed inclusive capacity-building practice while co-producing essential knowledge components integrated into project development (see Section 6).

In this UCCRN Element, capacity building is defined as "the process by which individuals, groups, organizations, institutions, and societies increase their abilities to perform core functions, solve problems, define and achieve objectives, and understand and deal with development needs in a broad context and in a sustainable manner" (Milèn, 2001). For climate change, it is important to develop capacity in both developing and developed countries. Climate change is global in nature and building capacity for both mitigation and adaptation in all nations will enable smoother international cooperation. Building capacity to expand the green jobs sector in both developing and developed countries will also provide mutual, symbiotic economic opportunities and resiliency.

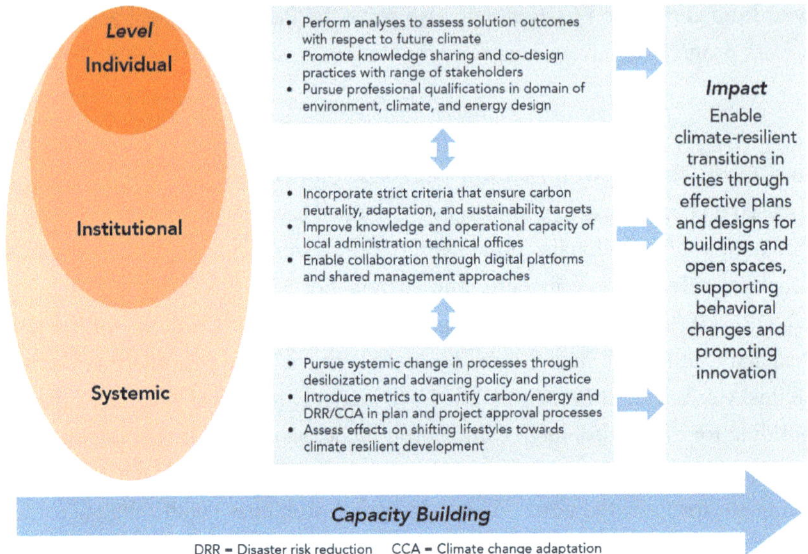

Figure 16 Climate change capacity building at the individual, institutional, and systemic levels with a focus on urban decision-makers and practitioners (Adapted from UNFCCC, 2023).

Following the UNDP (1998), the UN (2023a, 2023b) adopted a three-level approach in its climate change capacity-building activities: individuals, organizations and systems (see Figure 16). When assessing the UN's framework on climate change capacity, Figure 16 articulates the development of knowledge, skills, and competency across the three levels to achieve climate change outcomes. Competency is defined by the European Commission (2008) as: "the proven ability to use knowledge, skills, and personal, social, and/or methodological abilities in work or study situations and in professional and personal development."

Effective climate change action needs more than the acquisition of knowledge and skills, but the development of competence. Competence and functional skills are recognized components of being a professional (Biggs & Tang, 2007). In this context, professional, project, and process-based agency of planners and designers is expanded through creativity, multidisciplinarity, systems thinking and co-design (see Section 3). This is key to building the capacities of colleagues, clients, community stakeholders, and institutions to change attitudes and behaviors at all levels to stimulate proactive and effective climate action across sectors (Murtagh & Sergeeva, 2021; Arnett, 2023).

7.2 Enhancing Climate Change Capacity for Built Environment Professionals

This section's focus is climate change capacity building for built environment professionals (with a focus on urban planners, designers, and architects), and the institutions and systems within which they operate. Both developing and developed nations are considered. The actors involved in the production of the built environment include practitioners, professionals, academics, researchers, policymakers and community members. These actors vary across the life stages of producing and altering the built environment. They come from disciplines including urban planning, design, architecture, landscape design, construction, and engineering, and can also be from secondary sectors and actors who have a role supporting the production of the built environment (Hartenberger et al., 2013).

Each sector and actor brings with them skill sets and capabilities that contribute to different components of the built environment and the regulatory and policy frameworks that guide its development. Moving from the status quo to facilitating the necessary climate change actions in each of these sectors/across multiple actors is critical. The built environment process encompasses multiple interrelationships, and thus the need for urban planners, designers, and architects to work with actors from other sectors and key stakeholders to advance climate change action.

Four broad challenges to facilitating climate change action in built environments relating directly to capacity building are identified:

- *Need to increase the number of built environment professionals in practice.* A recent collaborative study by the Commonwealth Association of Architects, the Commonwealth Association of Planners, the Commonwealth Association of Surveying and Land Economy, and the Commonwealth Engineers Council (2020) found a critical shortage of qualified architects, urban planners, engineers and surveyors in many developing Commonwealth countries, demonstrating the need to both increase the number of architecture, urban design, and planning professionals working in cities, as well as enhancing climate training for existing professionals.
- *Need to build the climate change capacity of existing professionals.* Recent research indicates that there are climate change capacity gaps expressed by built environment professionals and urban planners in Australia and Canada, and coastal managers in the US (Tribbia & Moser, 2008; Canadian Planning Institute, 2019; Hürlimann et al., 2023a, 2023b). Similar skills shortages were explored in depth by the UK Parliamentary House of Commons in relation to built environment professionals since 2007, highlighting a lack of clear

strategy, inadequate supply of skilled workers, and low levels of diversity in the existing workforce, especially in relation to the current "Net Zero" Greenhouse Gas Emissions Policy (UK House of Commons, 2008; UK House of Commons, 2021; UK Parliamentary Committee on Climate Change, 2023). The EU also finds this labor shortage more generally in construction and engineering, particularly in relation to Net-Zero skills (European Commission, 2023).

- *Need to develop climate change curriculum in education programs degrees*. Recent research into the coverage of climate change in urban planning degree curriculums across multiple contexts has shown that climate change is not well addressed (Preston-Jones, 2020, Hürlimann et al., 2021c). Collaborative efforts across fields are seeking to address curriculum gaps. For example, Planners for Climate Action (P4CA 2023a,b) have developed a repository of course manuals with the explicit purpose to facilitate climate change capacity building for urban planners. In 2023, UCCRN and MIT delivered a free online course on "Cities and Climate Change," focusing on core interdisciplinary topics supporting urban climate mitigation and adaptation. Yet, further coordinated work is needed to advance curriculum and continuing professional development opportunities.

- *Need to complement mainstream approaches by engaging a diverse actors*. This includes Indigenous peoples, gender-specific constituencies and communities, using transdisciplinary and holistic approaches and co-creating pathways that gives meaningful space for marginalized actors to participate in, direct, and benefit from the capacity-building agenda (Klinsky, 2022) (see Box 5).

BOX 5 SOCIAL TECHNOLOGY: SOLUTIONS TO EMPOWER
VULNERABLE COMMUNITIES

Social technology is defined as reapplicable products, techniques, or methodologies, developed with the community and appropriated by it, which represent effective solutions for social transformation to improve living conditions and social inclusion (ITS, 2004; Dagnino, 2010; Pozzebon, 2015). Technology can be classified as social when it proposes to act on a social problem and its values are informed by the development of society, not the market (Neder, 2011; Pozzebon, 2015; Addor, 2021). They are alternative technologies to conventional technology, sustainable and low-cost, and appropriate to the principles of

Box 5 (cont.)

solidarity economy and social justice (Dagnino, 2014). The social technology movement gained strength at the beginning of the twenty-first century in Brazil and continues to grow (Dagnino, 2014; ITS, 2004; Pozzebon, 2015).

Technology is inseparable from the individual's culture in social technology (Neder, 2011; Pozzebon, 2015). The incorporation of individual and community knowledge and their interaction with technicians and researchers involved is decisive for social practice. This interaction has three inseparable principles: the formative experience formation through the day-to-day experience of social individuals as learning and training; technological culture – treated as a process of sociotechnical adequacy; and the self-organization of social individuals understood as a space for constructing appropriate self management methods by the social groups involved (Neder, 2011).

Urban adaptation and mitigation solutions based on the field of social technology promote justice and increase the resilience of marginalized communities' response to climate crises by placing people at the center of solutions and incorporating inclusive, emancipatory and empowerment processes, and reformulate relationships of power (Heimann & Mallick, 2016; Addor, 2021). Social technology can be identified in solutions for climate change adaptation and are often included in urban planning and design actions, such as for food security, like the urban agroecological practices developed in urban peripheries and slums in São Paulo and Rio de Janeiro (Levidow et al., 2021), in urban co-design such as those developed in Brazilian favelas to create urban gardens (Montuori et al., 2017); or for disaster risk reduction (DRR), such as community-based early warning systems (EWS), as in the cases for floods and landslides in São Paulo, Brazil (Marchezini et al., 2017).

7.3 How to Build Capacity

Capacity building typically follows three phases in a continuing cycle: assessing needs; developing strategies to address these needs; and monitoring and evaluating the capacity-building actions (Milèn, 2001). Critical capacity gaps have been assessed in a context-specific way across the three scales of individual, organizational, and systemic – for each built environment sector and its actors. However, limited assessment of climate change capacity gaps and needs has been undertaken for built environment professionals and the sector at large.

7.3.1 Developing Strategies to Address Climate Change Capacity Needs

There is an increasing body of work conducted by built environment professional associations and independent organizations at both international and national scales to advance climate change capacity across sectors of the built environment. Impetus for many initiatives appears to have come from the Paris Agreement (UNFCCC, 2015) and IPCC reports, including the Special Report on 1.5°C (IPCC, 2018). In response, some professional organizations at national and international levels made declarations of climate change emergency (e.g., IFLA, 2019; RIBA, 2019) and have facilitated actions to improve sector competency. (See Additional Resources, Figure 9, for a Sankey diagram of climate change facilitators for built environment professionals.)

Information sources trusted by built environment professionals include industry bodies, government sources, and academic researchers. These results were found in the Australian property and construction sectors, and in urban planning studies (Hürlimann et al., 2018; Canadian Institute of Planners, 2019; Warren-Myers et al., 2020). Research also indicates that colleagues can be an important source of climate change information for coastal managers and urban planners (Tribbia & Moser, 2008; Canadian Institute of Planners, 2019; Hürlimann et al., 2023a, 2023b). However, work is uneven across sectors and among professional organizations and is not necessarily translated into current university curricula.

Capacity building for climate change in built environment professions can encompass a range of formats and structures. Formal programs include degrees or diplomas provided through universities, of which some may receive professional accreditation. Less formal opportunities include short courses or training modules facilitated by professional associations or government authorities. Similarly, professional association publications and resources can be important catalysts and guides for change. Capacities also can be built through new "Learning Landscapes" centered on 1. Integrating pedagogical tools to generate individualized learning experiences; 2. Enabling educators to conceptualize sociocultural influences in and beyond educational space boundaries; and 3. Specifically designing spaces for experiential learning (Hansen, 2012; Neary et al., 2010) (see Case Study 6). (See Additional Resources for specific capacity building tools.)

7.3.2 Monitoring and Evaluating Capacity-Building Actions

Monitoring and evaluating the success of capacity-building actions is critical to improving future processes (Milèn, 2001). Clear articulation of the purpose of monitoring and evaluation activities is useful to increase effectiveness. Milèn (2021) identifies some current concerns in the monitoring and evaluating of capacity-building actions, particularly from a developing country context. More emphasis is needed on measuring processes rather than results, for example, evaluating the capacity building itself as well as adaptation outcomes and understanding effectiveness of partnerships over time.

CASE STUDY 6 FLOODING ADAPTATION STRATEGY IN APARTADÓ'S RIVER MASTER PLAN[23]

Sara Arteaga-Morales and Juliana Vélez-Duque

Apartadó, Colombia, is located along the Apartadó River in the Urabá region and is home to one of the thirty-six recognized biodiversity hotspots around the world, the Tumbes-Chocó-Magdalena in Colombia. Severe flash flooding occurs in the river basin approximately every six to ten years, exposing almost half of the city of Apartadó and its neighborhoods, which include seventy-two Indigenous and six afro-Colombian communities, to damage and displacement. With precipitation projected to increase due to climate change, extreme weather events associated with the river will become stronger and more frequent. Recognizing the need to adapt to climate change, the Apartadó River Master Plan aims to reduce risk exposure and increase flooding protection along the river basin through participatory planning, nature-based solutions, and ecosystem restoration.

The methodology for the Master Plan included three stages: due diligence, diagnostics, and formulation. Community mapping and participatory planning workshops were conducted with different communities to understand their relationship with the river at cultural and economic levels their knowledge of ecological restoration, and to build their capacity for experiential learning. The plan's formulation stage resulted in proposals that encompassed environmental education, disaster preparedness, ecological restoration, community monitoring, eco-resilient neighborhoods, and nature-based tourism. It also emphasized preservation of cultural and ecological memory associated with the river.

---

[23] See extended version of case study at https://uccrn.ei.columbia.edu/case-studies.

Figure 17 Community mapping and participatory planning workshop during the development of the Apartadó River Master Plan. (Source: Sara Arteaga, 2023a)

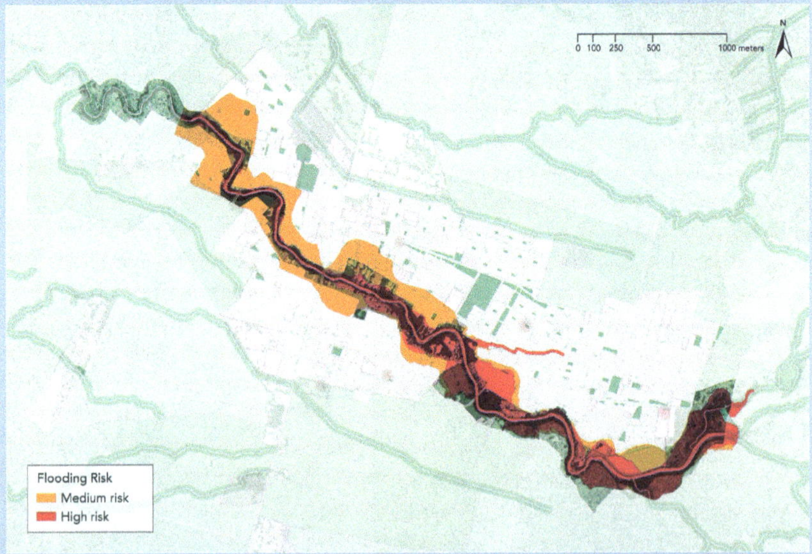

Figure 18 The Apartadó River crosses the city from east to west. (Source: Sara Arteaga, 2023b)

CASE STUDY 6 (cont.)

The Apartadó River Master Plan addresses risk management, governance issues, environmental degradation, social inclusion, and cultural representation to create a proposal that strengthens disaster response within the city. It promotes climate and environmental justice, cultural appropriation programs, and public–private partnerships to tackle weak governance institutions. The methodology demonstrates the importance of centering adaptation interventions around communities and local stakeholders that can guide the narrative of disaster response while incorporating unique cultural elements to risk management. This climate and environmental justice approach promotes social empowerment within communities and enables decision-support frameworks and resilience.

8 Metrics, Performance Indicators, and Tools

Analytical tools and guidelines are essential for supporting urban planning, design, and architecture for mitigation and adaptation. By means of scientific data and corresponding metrics analyzed by such tools as evidence, performance of the built environment can be assessed quantitatively thus providing a knowledge base for action.

Design guidelines are defined as sets of practical recommendations or codes on how to design strategies, plan actions, and apply measures to respond to climate challenges. Design guidelines coalesce multidisciplinary knowledge to advance climate action. They are used by planners and policymakers to determine how to implement principles efficiently and meet their purpose appropriately. Both analytical tools and guidelines are primarily stratified by the spatial scale on which the research or application is focused. (See Additional Resources for selected guidelines developed in recent years.)

Technologies such as geospatial mapping, a technique that creates customized maps by displaying spatial data in a geographic context, and parametric design, a technique where engineers use algorithms to create complex structures or products, often engage multiple scales, community jurisdictions, stakeholders, and governance. Drawing from planning, urban design, and building scale experience, a current theme emerging from the use of these technologies is "silo-busting" across spatial scales and sectors.

Such an approach highlights the importance of addressing the climate-resilient urban transformation process through a multiscale perspective to enhance the coherence of concepts, methods, and assessments at different urban planning and design scales (i.e., natural areas, building, neighborhood, city, and metropolitan region) (Leone & Raven, 2018). On the other hand, scale-dependent specificities need to be considered to effectively incorporate principles, models, tools, and design priorities within planning and urban design for specific projects, also in relation to sectoral regulations and technical specifications.

The purpose of this section is to discuss the principles relevant to different spatial scales in urban areas as they relate to mitigation and adaptation and to clarify the issue of spatial scales in analytical tools and guidelines. Main metrics and indicators are introduced for evaluating GHG emissions reduction and climate-adaptive performance of the urban environment. State-of-art tools for analyzing urban environments, including models, frameworks, software, and representative guidelines for regulating design practice are reviewed on a case-by-case basis.

8.1 Climate Scales versus Planning and Urban Design Scales

In urban climate science, a useful hierarchy of scales correspond to different dominant processes that inform observation and modeling (Oke et al., 2017). The background climate/weather within which metropolitan regions are embedded operates at the regional or mesoscale (~500 km). The influence of cities on weather through modifications of land cover and subsequent impacts on the overlying air occurs at an urban scale (~10 km).

At the local or neighborhood scale (~1–5 km), the influence of distinct urban land use land cover (LULC) types (e.g., residential, commercial, green space) is evident; these are linked with typical building dimensions, tree canopy cover, and traffic patterns (Chow & Roth, 2006; Chow et al., 2012). Each neighborhood type has a myriad of climates at the microscale (≤100 m) that are created by the specific characteristics of buildings (e.g., dimensions), layout (e.g., street width and orientation), and landscaping. The range of these microclimates is consistent across a neighborhood type.

Using a similar spatial framework, planning, urban design, and architecture practice can be categorized by scale from regional plan to building design. The hierarchies of spatial scales in these two working systems can be correlated, as shown in Figure 19. Thus, both communities of practice can understand each other reciprocally and transfer knowledge effectively.

At the mesoscale to urban (city) scale, urban climate research provides background information, including climate change projections.[24] The climate models that operate at this scale do not routinely include urban land use types but rather simple urban land cover. The outputs of these models correspond to the scale of synoptic weather reports and observations of weather patterns over large areas at specific times. They can be used to support planners and policymakers as they

---

[24] See ARC3.3 Element on *Urban Climate Science: Knowledge Base for City Risk Assessments and Resilience*, www.cambridge.org/core/publications/elements/elements-in-climate-change-and-cities

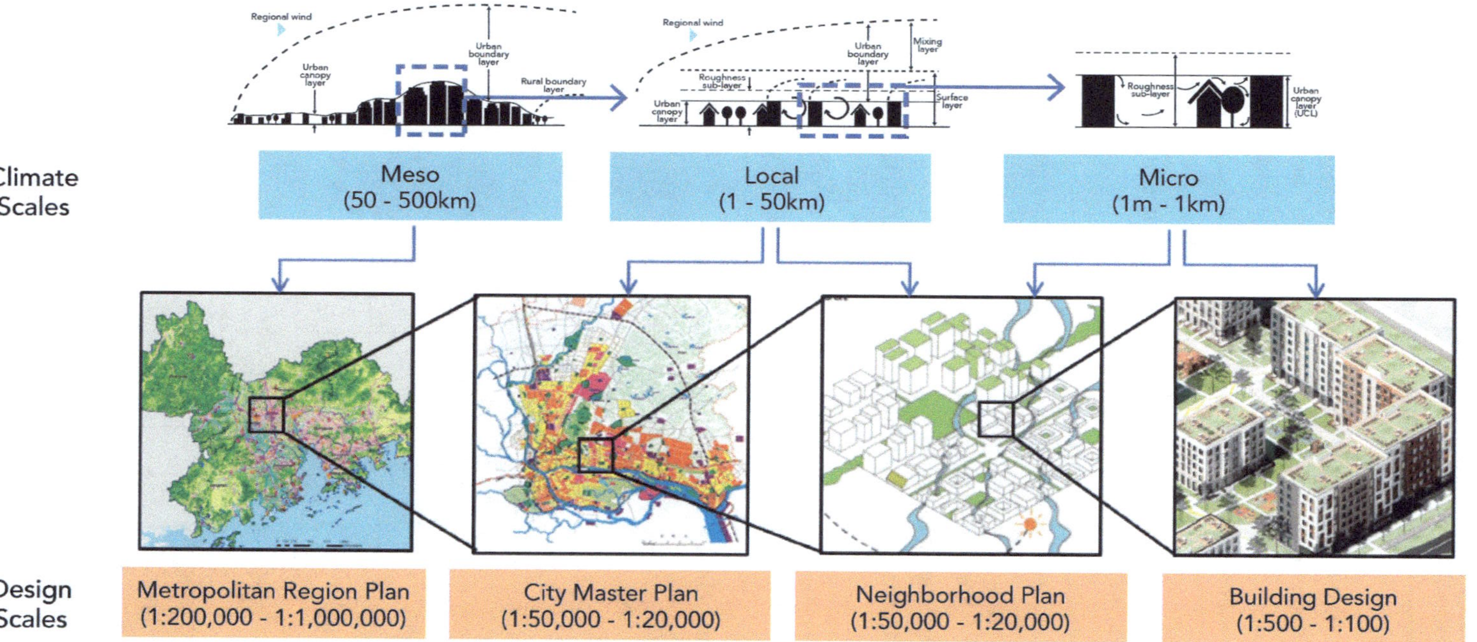

Figure 19 Relation between urban climate scales and design scales.

prepare regional plans to reduce long-distance travel related to high GHG emissions and propose design guidelines for responding to extreme events such as coastal flooding.

Global analysis data include weather observations, computer simulations of historical weather patterns, and climate model projections. They describe current and future climate, including extreme events such as heat waves and heavy downpours, providing a context for urban-scale decisions (see https://cds.climate.copernicus.eu). Climate models that focus on city and neighborhood scales use "urbanized" models that can account for different urban LULC types and three-dimensional form (morphology).

Observational weather data at city and neighborhood scales is not generally available, although there are a growing number of cities that have their own weather station networks or have utilized volunteer citizen weather station data that capture the UHI effect. New York City, London, Berlin, and Nairobi are all cities that use both independent weather stations and volunteer citizen weather data (Klimastadt Berlin, 2023; NOAA, 2025; TAHMO, 2024; UK Met Office, 2011; USDA, 2024). This scale corresponds to that of urban planning (i.e., cities and neighborhoods) related to decisions on land use, transportation, and urban morphology that can enhance climate resilience. Correlating climatic and policy information via the integration of spatial scales is essential for better understanding the urban environment and proposing climate-aligned architecture, design, and planning actions (see Case Study 7).

CASE STUDY 7 INTEGRATED STORM WATER MANAGEMENT IN NEW YORK CITY: IMPLEMENTATION OF GREEN INFRASTRUCTURE MEASURES[25]

Martina Kohler

Every year in New York City, around 20 billion gallons of untreated raw sewage and polluted runoff are diverted from the City's wastewater treatment plants during combined sewer overflow (CSO) events (Levine, 2020). These are directed into the rivers along the shoreline of the city because the designed capacity of the system is reached quickly when it rains. Future climate change-induced increases in precipitation will put further stresses on the already overloaded system. This case study assesses the implementation of NYC's Green Infrastructure (GI) plan to reduce rainwater overflow and address combined sewer overflows to both adapt to increased rainfall and to reduce Urban Heat Island (UHI) effects.

---

[25] See extended version of case study at https://uccrn.ei.columbia.edu/case-studies.

New York City's GI program aims to reduce combined sewer overflows into NYC waterways and the UHI effect through urban greening. The plan, introduced in 2010, has the overarching goal of enabling NYC to manage 1" of storm water runoff with GI across 10 percent of the impervious surfaces within the combined sewer area of the city by 2030. More than 11,050 GI assets, predominantly rain gardens, have been installed or are currently under construction since 2012.

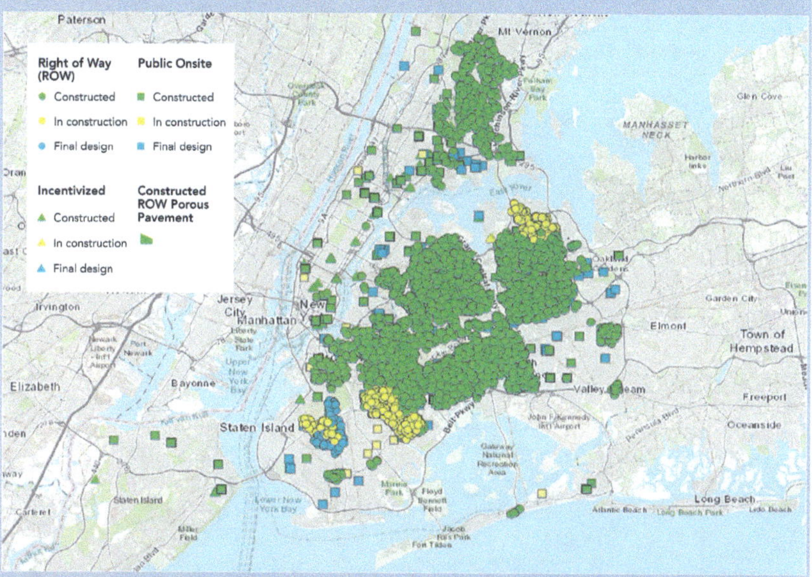

Figure 20 New York City Department of Environmental Protection (DEP), Green Infrastructure Program Map.
(Source: NYC OpenData, State of NJ, Esri, HERE, Garmin, USGS, NGA, EPA, NPS, April 13, 2022)

The installation of rain gardens in the public right of way faces some challenges including necessary maintenance to ensure proper performance, conflict with density of city infrastructure below the streets and sidewalk surfaces (gas, cable, freshwater lines), high bedrock, and high-water-table levels in some areas of the city.

The GI program has made progress despite unanticipated challenges. Reaching its stated goal in 2030 will require an increased focus on GI measures on private properties. Currently DEP's grant program and DEP's Private Property Retrofit program both focus on private properties 50,000 square feet and larger. As a result, there remains significant unrealized potential to capture water and divert roof runoff from the sewer system in many smaller catchment areas.

Figure 21 Rain garden installations at Denton Place, Brooklyn, New York City, part of the Green Infrastructure Plan.
(Source: Kohler, 2023)

8.2 Analytical and Modeling Tools for Climate Mitigation and Adaptation

To implement climate-aligned interventions, numerous analytical and modeling tools have been developed to support the work of researchers and practitioners. These are relevant to urban scales and aid in assessing the effects of planning and urban design scenarios in terms of climate mitigation and adaptation (Figure 22).

8.2.1 Analytical and Modeling Tools for Climate Mitigation

A greenhouse gas (GHG) inventory is fundamental for accounting GHG emissions in cities. The IPCC Guidelines on National Greenhouse Gas Inventories provide methods and rules for countries around the world to establish national GHG inventories and reduce emissions in four sectors: energy; industrial processes and product use; agriculture, forestry and other land use; and waste. At the city scale, the IPCC national source-based emissions accounting is not readily applicable. Therefore, over the last two decades, leading organizations and city networks have developed accounting protocols that focus on cities, including the International Standard for Determining GHG Emissions for

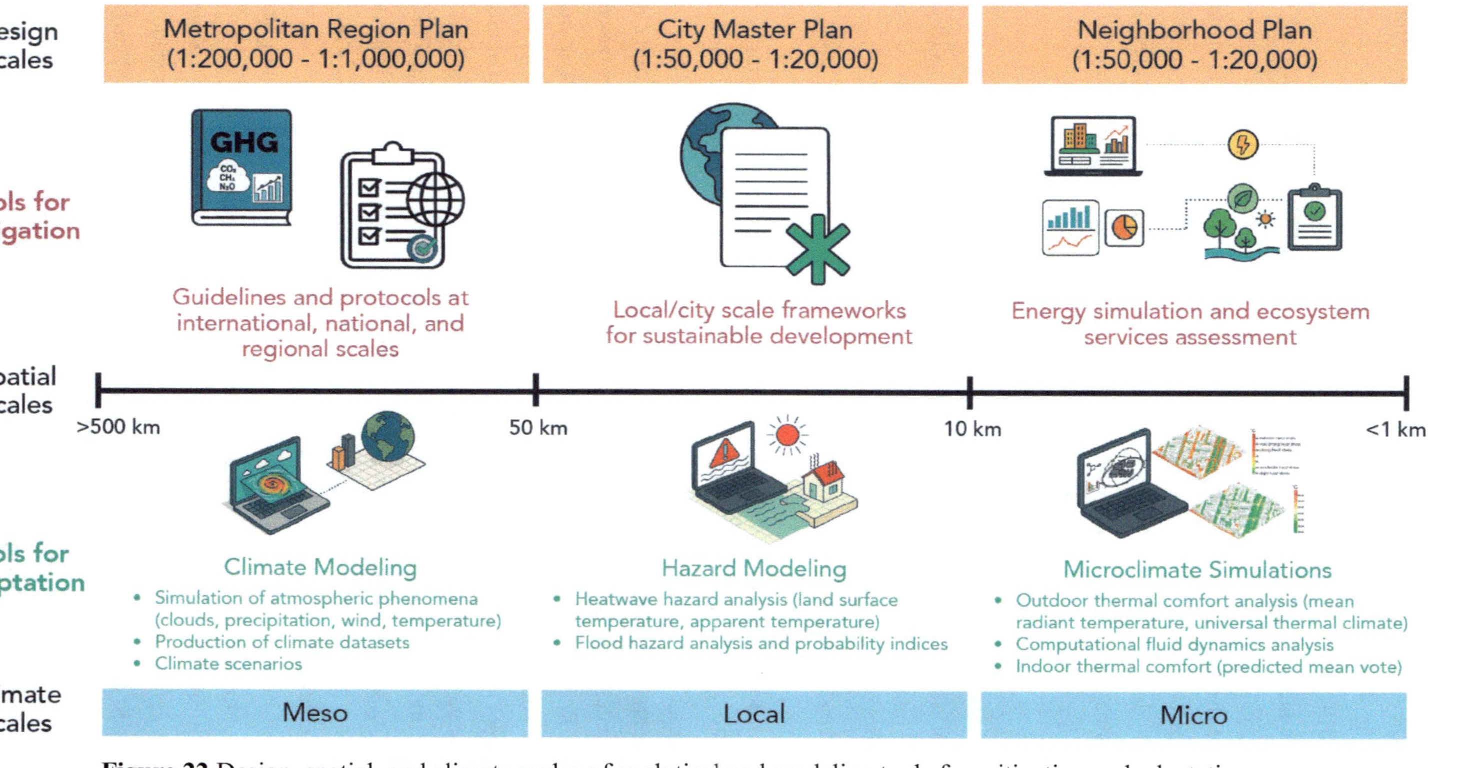

Figure 22 Design, spatial, and climate scales of analytical and modeling tools for mitigation and adaptation.

Cities, the Global Protocol for Community-Scale Greenhouse Gas Emission Inventories (GPC), and the US Community Protocol (UNEP et al., 2010; ICLEI USA, 2019; Fong et al., 2021, Davey, 2025).

At the building scale, building energy simulation software such as CEA,[26] DeST,[27] EnergyPlus, DOE-2[28] calculate building energy performance, and facilitate design to meet building energy conservation standards. Peking and Tsinghua University Civil Engineering Software (PKPM-CES) is a carbon emission design and analysis software applicable to building life cycles. It includes carbon emission analysis of design, construction, production, transportation, operation and maintenance, and demolition.

8.2.2 Analytical and Modeling Tools for Climate Adaptation

For adaptation to manage risks at the mesoscale and local scale, the Crichton's Risk Triangle framework (Crichton, 1999) can be linked with the risk concepts of the IPCC (IPCC, 2014). These methods provide useful tools to assess and map climate-related risks, such as flood hazards (Chen, 2021), extreme heat (Hua et al., 2021), and air pollution (Shi et al., 2020). The techniques overlay the frequency and intensity of hazards with the presence of exposed assets, thereby measuring the propensity or predisposition of urban areas to be adversely affected. (See Additional Resources, Figure 12, for a spatial distribution of heat vulnerability in Hong Kong.)

There are several mesoscale climate models capable of simulating weather at urban scales. The best known of these is the open-source community-based *Weather Research and Forecasting (WRF) model*. The urbanized version of this model accounts for variations in the urban landscape in the form of numerical descriptors (known as urban canopy parameters or UCPs). The resolution of this model is variable, but it has been applied at sub-kilometer scales (Chen et al., 2011). Running mesoscale models such as WRF requires a considerable amount of training and resources, beyond the scope of urban planning skills. Planners and urban designers can work with urban climate scientists to test relevant climate change scenarios and interventions for individual urban areas.

One of the difficulties in applying mesoscale models to study urban phenomena is acquiring the numerical description (UCPs) of the city's LULC composition. One approach to acquiring these is to use the *Local Climate Zone (LCZ) scheme*, which categorizes the urban landscape into ten neighborhood types, each of which is associated with a range of UCP values for climatically relevant variables, such as the impermeable surface fraction, UHI, and Sky View Factor (SVF) (Han et al., 2024). Recently, a global LCZ map has been generated that provides a basic physical geography of cities, which can be

---

[26] City Energy Analyst.   [27] Designer Simulation Toolkit.   [28] Department of Energy 2.

used in urban climate modeling. An online LCZ generator enables users to create their own city map of neighborhood types.[29]

For investigating climate conditions within urban areas, there are a range of models and techniques that can be employed to guide climate-sensitive planning. For example, the *Urban Multi-Scale Environmental Predictor (UMEP)* combines models and tools for climate simulations and is linked to the Quantum GIS system (Lindberg et al., 2018). The UMEP has a suite of submodels that can be used to examine surface and air temperatures, shadowing patterns and runoff and supports scenario testing.

A more intuitive and city-specific technique uses the *Urban Climatic Map (UC-Map)* method as an evaluation tool to integrate urban climatic factors and town planning considerations. It presents individual climatic phenomena and hazards on a common spatial frame and uses maps to combine planning relevant information (Ren et al., 2011, see Section 8). A complete UC-Map system is composed of an Urban Climatic Analysis Map (UC-AnMap) and an Urban Climatic Planning Recommendation Map (UC-ReMap) (Figure 23). The latter is a planning and action-oriented assessment base that can be operated at the city or the district scale. Currently, more than fifteen countries have developed their own UC-Map system and applied it to develop climatic measures or guidelines for local planning. (See Additional Resources for tools that utilize computational fluid dynamics.)

These tools have various limitations regarding spatial scale, modeling resolution, simulation time, and suitability. Therefore, the integration of multiple tools and simulations across spatial scales can be useful in responding to a range of needs in practice. For example, synthesizing the results of the UC-Map and LCZ facilitates

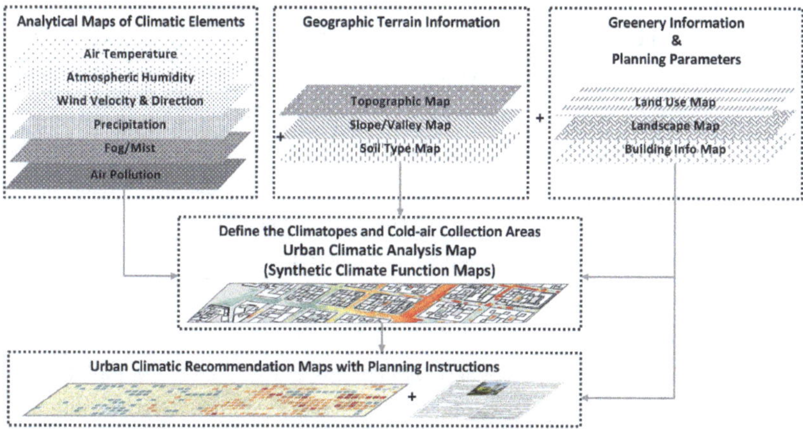

Figure 23 Structure of Urban Climatic Map (elaborated from: Ren et al., 2010).

---

[29] (https://lcz-generator.rub.de/).

the synthesis of building morphology and urban climate zones. This provides the evidence base for urban climate planning recommendations to support planners and decision-makers on improving mitigation and adaptation performance (Yin et al., 2022). (See Additional Resources for description of modeling tools to respond to extreme heat; see also Heat Vulnerability Index (HVI) map of New York.)

8.3 Metrics and Performance Indicators for Climate Mitigation and Adaptation

Mitigation

Planning, urban design, and building design practitioners can engage with the concepts and practices used to determine and assess GHG emissions. Global greenhouse gas performance is generally assessed at a smaller scale by the emission quantities (tCO_2) expressed by area (m_2); or total amount per capita or per gross domestic product (GDP) at a larger scale. ISO 37120 (2018) proposes that GHG emissions of cities be measured on a per capita basis, which represents the relationship between carbon emissions and individual consumption levels. GHG emissions per unit of GDP is a metric that shows the GHG intensity of the economy, commonly used in China.

For example, at the 2009 UNFCCC COP in Copenhagen, China committed to reducing the economic carbon intensity (i.e., CO_2 emissions per unit of GDP) by 40–45 percent from 2005 levels by 2020 (Cansino et al., 2015). The indicator was then distributed to provinces and cities, becoming a significant carbon performance measurement at every administrative level. The current carbon intensity reduction commitment of China is to reduce CO_2 emissions per unit of GDP by 60–65 percent from 2005 levels by 2030 (UNFCCC, 2020).

Improving energy efficiency and optimizing energy supply and distribution infrastructure are fundamental ways to reduce urban carbon emissions. The integration of renewable energy sources and energy recovery technologies such as utilization of waste heat contribute to lower the energy system emissions. Stockholm has been a leader in the utilization of waste heat, which uses the waste heat of data centers to power 2,500 residential apartments (Yuan et al., 2022). Indicators for non-fossil energy use and energy efficiency are essential to assess the performance for climate mitigation.

This growing momentum is evident in the actions of major cities, such as New York, London, Singapore, Paris, Tokyo, and Berlin, which are implementing both carbon metrics and energy-related indicators in climate change mitigation initiatives (GCoM, 2022). Smaller to medium-sized cities are also making robust climate commitments, with some of these cities proposing climate neutrality, reducing emissions through climate action to ensure no net effect on the climate system, or zero carbon goals, which aim to eliminate GHGs (Table 7).

Table 7 Climate action performance indicators of eight case study cities.

City	Area (km²)[30]	Baseline Year Emissions	Climate Mitigation Initiative	Launch Date	Performance Indicators
Berlin	891.85	1990	Climate-Neutral Berlin 2050 Berlin Energy and Climate Protection Programme 2030 (BEK)	2016 2018	CO_2 emissions reduced by at least 40% by 2020; by 2030, CO_2 emissions reduced by at least 60%; senate and administrative departments achieve carbon neutrality and eliminate use of coal, and most of BEK2030 will be implemented; achieve climate neutrality by 2050
London	1579	1990	1.5°C Compatible Climate Action Plan	2018	CO_2 emissions reduced by 60% by 2030, nearly 80% by 2040; zero-carbon by 2050;
New York City	789	1990	OneNYC 2050 Building a Strong and Fair City-A Livable Climate	2019	by 2040, share of clean energy in electricity to reach 100%; by 2050, 100% of GHG emissions eliminated, reduced, or offset; carbon neutrality achieved

[30] Area of city limits.

Table 7 (cont.)

City	Area (km²)	Baseline Year Emissions	Climate Mitigation Initiative	Launch Date	Performance Indicators
Paris	105.4	2004	Paris Climate Action Plan-Towards a Carbon Neutral City and 100& Renewable Energy	2018	CO_2 emissions reduced by 25% by 2020; by 2050, carbon emissions reduced by 75%, local carbon emissions reduced by 100%, energy consumption reduced by 50%, and energy consumption from 100% renewable resources
Singapore	719.9	2005	Taking Actions Today for a Carbon-efficient Singapore	2016	CO_2 emissions reduced by 36% and reach peak by 2030
Tokyo	2188	2000	Tokyo Climate Change Strategy	2011	With 2000 as the base year, GHG emissions reduced by 25% by 2020
Christchurch	1415.47	2016	Ōtautahi Christchurch Climate Resilience Strategy	2021	Net zero GHG emissions by 2045, halving emissions by 2030
Mérida	858.41	2000	Municipal Climate Action Plan	2018	Reduce emissions by 30% compared to the baseline in 2020, and by 2050, reduce emissions by 50% with 2000 as the base year

Adaptation

Human comfort and climate-responsive performance in urban spaces are predominantly evaluated through monitoring of thermal and wind environments. Urban climate depends significantly on the characteristics of urban morphology. Thus, urban morphology affects the corresponding metrics and indicators of thermal and wind environments.

Thermal environment

Air and surface temperatures are employed as the basic climatic variable for performance metrics to detect hot spots of heat risk in a city at the urban or neighborhood scales (Oke et al., 2017). Extreme event hours are accumulated as the very hot day hours and hot night hours during a certain period. In addition, human thermal comfort indices are widely used to represent human bioclimatic sensations and health risks due to the surrounding environment at the local or microscales.

The Universal Thermal Climate Index (UTCI), Physiological Equivalent Temperature (PET), Standard Effective Temperature (SET), Predicted Mean Vote (PMV), and Wet Bulb Globe Temperature (WBGT) are the most popular thermal indices adopted in urban and building design. The thresholds of the neutral values of these indices vary with climate zones, locations, and socioeconomic factors (Zare et al., 2018; Binarti et al., 2020; Jia et al., 2025).

The urban thermal environment is co-determined by urban morphological factors, including building density, vegetative cover, street canyon ratio (i.e., proportion between width of streets and height of flanking buildings), street orientation, sky view factor (i.e., the ability of built surfaces to emit radiation to the sky), and albedo (i.e., the ability of surfaces to reflect solar radiation). (See Additional Resources for a comparison of thermal perceptions in various bioclimatic indices.)

Wind

The wind environment is mainly assessed by wind speed and direction to represent urban ventilation performance. Urban ventilation can be severely reduced by poorly configured neighborhoods, resulting in extreme heat and frequent urban haze events. To increase the ventilation potential and improve quality of life within urban areas, designing neighborhood urban ventilation corridors is essential. Passive cooling strategies can be configured and implemented by regulating for climate-aligned urban morphology (Wai et al., 2025):

- The surface roughness in urban morphology is a key parameter that impacts wind direction and speed near the ground.
- Frontal area density is the ratio of a building facade to its total site surface area.

- Building coverage ratio is a metric that compares the area of a building to the size of the land it is built on.
- Floor area ratio (FAR) is a measurement of a building's floor area in relation to the size of the site.

8.4 Barriers and Bridges to Integrating Mitigation and Adaptation

Successful integration of carbon-neutral and climate-adaptive principles into urban design and planning processes faces three major challenges:

- Understanding of Scope 1, 2, and 3 GHG emissions[31] connected to critical urban systems such as energy, mobility, materials, food, water, and waste (Wiedmann et al., 2021; Lwasa et al., 2022);
- Management of complex information on climate impacts (IPCC, 2014); and
- consideration of the range of stakeholder interests and concerns (Rotter et al., 2013).

Taking up these challenges requires multiscale and multidisciplinary perspectives integrating economic, environmental, and social aspects, as well as dedicated stakeholder engagement. Holistic considerations of climate change adaptation have in recent years brought forward the concept of *adaptation cobenefits*, now embedded in the concept of *climate-resilient development* (Raymond et al., 2017; IPCC, 2023). Examples of relevant co-benefits linked to climate action in cities include increased quality of public spaces and social services; employment and income generation from new green jobs creation; improved quality of water, soil and air; increased urban biodiversity (Bachra et al., 2020). Novel climate-resilient planning and urban design tools connect climate benefits in terms of mitigation and adaptation with relevant social, economic, and environmental co-benefits responding to human needs and enhancing quality of life of urban communities (see Sections 6, 8, and 9).

9 Urban Design Climate Workshops (UDCWs)

The Urban Design Climate Workshop (UDCW) process builds upon established paradigms of contemporary planning, urban design, and architecture through

---

[31] Scope 1 refers to direct emissions that are owned or controlled by the organization. Scope 2 encompasses indirect emissions resulting from the consumption of electricity, steam, heating, and cooling. Scope 3 covers all other indirect emissions associated with the organization's upstream and downstream activities (World Resources Institute (WRI) and World Business Council for Sustainable Development (WBCSD), 2004).

the lens of a resilient and sustainable built environment.[32] It views the city as a body of work made by multiple actors; an indivisible and evolving organism shaped by infrastructure, development, public process, and density. To bridge climate science and climate action, policymakers, researchers, practitioners, and stakeholders need planning and urban design strategies to identify, configure, and evaluate urban climate factors at a neighborhood scale.

UDCW participants develop scenarios and urban prototypes aligned with climate resilient, net-zero carbon principles to strengthen community adaptability to climate change, reduce energy consumption in the built environment and enhance the quality of the public realm. With UDCW sites in diverse global cities, the planning process works to balance bold visions for the city with marginalized communities' access to services. As UDCWs rely on engagement and co-design, these participatory workshops also act as stakeholder sounding sessions that occur throughout the ARC3.3 assessment process.

The UDCW is conceived as a hands-on, capacity-building process that engages urban designers, climate scientists, policymakers, students, and stakeholders. Working side-by-side, this cross-sectoral planning process envisions how urban design and planning can shape transformative climate action in urban districts. Scenario modeling illustrates likely climate impacts from development and rezoning alternatives, while climate-sensitive prototyping identifies opportunities for GHG mitigation and climate adaptation. UDCWs configure interconnected microclimates and urban systems within the city to achieve reduced energy loads, cleaner air, and enhanced civic life, while incorporating mitigation and adaptation. Since 2015, a series of UDCWs have taken place in New York, Paris, Naples, Durban, Randers, and Rio de Janeiro. (See Additional Resources for UDCWs in Paris, Durban, Naples, and Sunnyside.)

Drawing from evidence-based urban climate factors (Figure 23), this collaborative planning and urban design process was summarized in the Urban Planning and Urban Design Chapter of the *Second Assessment Report* on *Climate Change and Cities* (Raven et al., 2018). Benefitting from additional years of UDCW experience, this ARC3.3 Element assesses research, field testing, and validation of urban systems, built environment models, and frameworks that integrate climate mitigation and adaptation. This integrated urban transformation reflects a value proposition paradigm shift that includes environmental justice, synergies between research and practice, innovative tools, capacity building, and clear roadmaps for climate action.

---

[32] The UDCW approach and toolkit has been developed, tested, and improved through several research projects in Europe and the US including H2020 CLARITY; H2020 ESPREssO; Erasmus+ UCCRN_edu; Horizon Europe KNOWING; Horizon Europe UP2030; National Science Foundation – Belmont Forum and RCN City-as-Lab + Supplement.

Notwithstanding effective use of prototyping or developing a representation of the idea or solution to test before a product is launched to demonstrate positive outcomes, the goal of UDCWs is to ultimately go beyond "demonstrator projects," to embed these evidence-based approaches in the standard planning and urban design process. This would foster a rapid evolution in the "business as usual" approach to planning and design, where equity and environmental justice go hand-in-hand with measurable impacts on climate co-benefits goals. Urban Design Climate Workshop methods and tools identify, configure, and evaluate responses to stakeholder priorities and urban climate factors through multiscale planning and design strategies and solutions.

9.1 UDCW Process and Phasing

The UDCW methodology focuses on sequential and iterative phases that lead to the development of neighborhood transformation through a multidisciplinary and multiscale approach (Raven et al., 2018). Phases of the methodology are implemented with the support of UCCRN multidisciplinary experts, community experts, and other stakeholders. This process combines knowledge sharing and co-design actions with urban decision-makers and local communities together with the development of simulations based on computational design tools to control the main indicators that determine the performance of buildings and open spaces in relation to climatic stress conditions. (See Additional Resources for a list of partners engaged by local city hosts and UCCRN Regional Hubs that helped to develop UDCWs.)

Depending on the location, objectives, available resources, and participants, different types of UDCWs have been designed. A key feature is that they are designed to be iterative and replicable, not "one-offs." Multiple encounters enable the development of long-term relationships with stakeholders and of persistent engagement with climate change challenges experienced by the participating city over time.

Urban Design Climate Workshop types can be summarized as follows:

- *Knowledge Exchange*. One to two days (plus two to four weeks preparatory activities involving UCCRN team) with panel discussion/post-it session with UCCRN facilitators, session, or side event within a conference (with experts, scientists, practitioners) (Example: Bonn UDCW, 2019).
- *Capacity Building*. Three to five days (plus four to eight weeks preparatory activities involving UCCRN team and local hosts), a workshop with multidisciplinary UCCRN experts, local authorities' representatives (technical experts, scientists, practitioners, local authorities and communities) (Example: Durban UDCW, 2019).
- *Design Studio*. Seven to fifteen days (plus eight to sixteen weeks preparatory activities involving UCCRN team and local hosts), a workshop with

multidisciplinary UCCRN experts, students (technical experts, scientists, practitioners, local stakeholders, and communities, with an opening conference and a final event involving external audience (Example: Gowanus UDCW, see Case Study 8; UCCRN_edu UDCW series, 2022–2024).

The UDCW process is intended to explore "climate synergies" that can be activated with respect to stakeholder priorities, specificities of urban systems, and planning/design opportunities in relation to the study area and the targeted program. The UDCW comprises stakeholder priorities, urban systems, co-design, Metrics, Policies and Feedback Loops. The UDCW Design Process (Figure 27) describes six phases:

1. Synergies & Adjacencies: Urban Systems & Users;
2. Synergies & Adjacencies: Spatial Scales;
3. Hybrid Neighborhood: Simulation & Modeling;
4. Urban Climate Factors;
5. Urban Form - Opportunities & Constraints;
6. Urban Climate Goals

These phases are not necessarily engaged in chronological order but are interchangeable, based on specific needs.

9.2 Urban Climate Factors

Urban Climate Factors have been updated in this Element following five years of UDCW activities (2018–2023), and represent a fundamental tool used throughout UDCW phases for planning and urban design (Raven et al., 2018; Raven, 2019) (Figure 24). Mitigation and adaptation solutions are embedded within a knowledge-sharing process to explicitly connect stakeholder priorities to climate benefits. This provides a framework to map co-benefits for climate action within the local context.

The four Urban Climate Factors are:

- *Efficiency of Urban Systems*: Efficient urban systems reduce anthropogenic emissions from energy, food, transportation, and waste. Reducing the impact of polluting vehicles, manufacturing, construction, and waste heat from buildings are critical priorities.
- *Form and Layout*: Climate-aligned urban design can configure densely occupied urban settlements to offset a challenging local climate context. Urban districts can be designed to enhance cooling and ventilation to reduce energy use and allow citizens to cope with higher surrounding temperatures, while enabling cities to better manage flooding. Climate-aligned urban design that integrates compact urban form with natural systems can achieve a network of attractive and healthy microclimates.

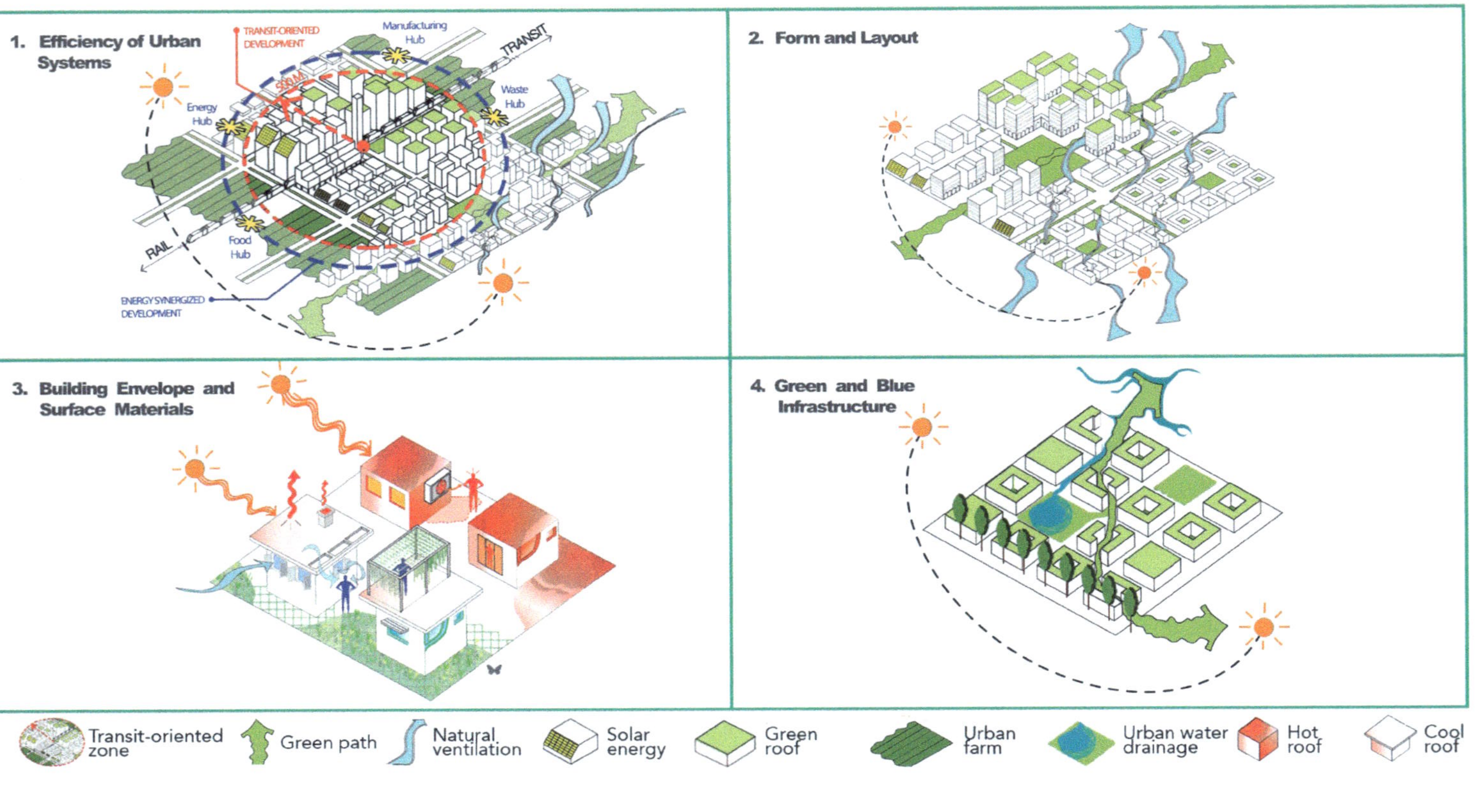

Figure 24 Urban Climate Factors
(Adapted from Raven, 2019)

- *Building Envelope and Surface Materials*: Selecting low heat capacity construction materials and reflective coatings can improve building performance by managing heat exchange at the surface. Air conditioning-reliant buildings are often isolated from their neighborhood microclimate. One approach is to define a wider range of acceptable indoor temperatures by enabling buildings to be better connected to healthier, outdoor microclimates, whose surface materials and equipment enhance shading, solar reflection, and reduce water infiltration
- *Green and Blue Infrastructure*: Increasing vegetative and tree canopy cover can simultaneously lower outdoor temperatures, building cooling demand, rain and floodwater runoff, and pollution, while sequestering carbon. Investments in pedestrian and cycling corridors aligned with parks, water bodies, and other natural systems, can reduce carbon emissions, enhance carbon sequestration, capture stormwater and perhaps most effectively, cool cities through evapotranspiration and shading.

Shaped by phased policy mandates and aspirations emerging from stakeholder priorities, the synergies between urban systems and climate play a central role in configuring the mitigation and adaptation design process. The urban systems include energy, food, water, waste, transportation, natural systems, and the built environment. The climate profile includes heat, flooding, drought, and GHG emissions.

The urban systems and climate profile form the basis for modeling future development scenarios. The models assess potential mitigation and adaptation outcomes based upon local climate projections by:

- Performing robust climate hazard/impact and mitigation/adaptation assessments based on IPCC climate scenario approaches
- Streamlining use of quantitative and qualitative indicators to support multi-scale evaluation of planning and design solutions across metropolitan region, city, neighborhood, and building scales
- Linking climate mitigation and adaptation outputs from urban development projects to social, economic and environmental co-benefits

Climate analysis maps identify urban zones subject to the greatest impacts associated with rising temperatures, increasing precipitation, and extreme weather events. A climate analysis map is based on spatial reference data. For climate adaptation, commonly employed climate analysis maps include urban heat "hotspots" and flood zone maps. The spatial resolution is tailored to the urban planning and urban design level.

Urban Design Climate Workshop climate analysis mapping is conducted at the district and subdistrict scale. District-scale mapping analyzes surface radiation obtained from LiDAR satellite data. These are a proxy for UHI and urban microclimate that helps identify district-level "hot spots," i.e., zones of greatest heat intensity. Potential pilot subdistricts are identified from this hotspot data. Priority districts characterized by population activities, projected development, and infrastructure investment overlap natural hazard zones and extreme weather events. In effect, as the urban climate gets hotter and wetter, higher density increases urban heat stress; and risk of flooding. Urban climate and hazard/impact assessment models and tools (e.g., Solweig, LadyBug, EnviMet, Sfincs, HEC-RAS, and the UDCW simulation tools – such as GIS tools, three-dimensional modeling, and climate-resilient neighborhood three-dimensional configurator) — can identify priority zones and hotspots at a neighborhood level. (See Additional Resources, Box 3, for a description of UDCW simulation tools.)

Supporting tools have been developed to analyze relevant metrics linked to the urban climate factors within identified priority areas (Table 8). Analytical

Table 8 Urban climate factors, indicators, and metrics.

Urban Climate Factors	Indicators	Metrics
Efficiency of Urban Systems	Stationary energy Waste Food Transport Manufacturing Waste heat Circular economy indicators GHG emissions per capita	Average building energy consumption (kWh/m^2y) Total energy consumption (kWh/m^2y) Renewable energy production potential (kWh/m^2y) GHG emissions (tCO_2e/yr) Hospitalization costs (\$) Mortality rate increase (%) Vehicle Kilometers Traveled (MKT) Municipal solid waste generation per capita
Form and Layout	Massing diagrams Wind-Sun impacts Sky view factor Land surface temperature Thermal comfort analysis Heat wave hazards Heat wave impacts	Mean Radiant Temperature (TMRT) (°C) Universal Thermal Climate Index (UTCI) (°C) Apparent temperature (°C) Land surface temperature (°C) Energy use intensity (kWh/m^2y)

Table 8 (cont.)

Urban Climate Factors	Indicators	Metrics
	Energy profile	Renewable energy production potential (kWh/y)
		Albedo (0–1)
		Computational Fluid Dynamics (CFD) (m/s)
Building Envelope and Surface Materials	Radiation analysis	District energy consumption (kWh/m²y)
	Building envelope rating	Building energy consumption (kWh/m²y)
	Thermal comfort analysis	Renewable energy production potential (kWh/y)
	Heat wave hazard	Carbon footprint of building materials (tCO₂e/yr)
	Heat wave Impact	Indoor Predicted Mean Vote (PMV)
	Energy profile	Mean Radiant Temperature (TMRT) (°C)
		Universal Thermal Climate Index (UTCI) (°C)
		Apparent temperature (°C)
		Run-off (-)
		R-Value (K.m2/W)
		Albedo (0–1)
Green and Blue Infrastructure	Tree canopy cover	Tree canopy cover %
	Biodiversity	City Biodiversity Index (CBI)
	Heat wave hazard	Carbon storage potential from vegetation (tCO₂e/yr)
	Heat wave impact	Run-off (-)
	Flood hazard	Flood probability index (-)
	Flood impact	Impervious surface index (−1, 1)
		Building damage (structure/content) ($)
		Road infrastructure damage (cleaning/repairing) ($)

parametric design tools have been integrated into the UDCW simulation toolkit as a way for planners and designers to easily analyze climate change projections and local conditions determined by the urban climate factors. Off-the-shelf mapping tools generate and illustrate urban systems and climate outcomes, including heat hazards, flood-prone zones, and GHG emissions. GIS models perform two-dimensional analyses, while three-dimensional computational design tools simulate the climate and energy drivers from neighborhood to building scale. (See Additional Resources, Box 3.)

Drawing from the UDCW taxonomy for input data requirements at neighborhood scales, urban development scenarios are evaluated through the lens of integrated climate adaptation and mitigation benefits. These quantitative climate indicators are supported by quantitative and qualitative assessment of social, economic and environmental co-benefits. A three-dimensional modeling configurator supports a co-design process between practitioners (planners, architects, and engineers) and local authorities (see the Additional Resources, Box 4). This provides insights into the mitigation and adaptation performance of design alternatives at the conceptual stage of project development (Nocerino & Leone, 2023; Nocerino & Leone, 2024).

9.3 Stakeholder Priorities

The stakeholder priorities phase provides a program framework between aspirational urban transformation scenarios and the daily lives of people, emphasizing interconnections between among climate, social, economic, and environmental co-benefits. This phase includes collaborative and participatory activities to assess the needs and expectations of stakeholders and local communities in an inclusive co-production process (see Case Study 8). Methods include collaborative mapping and co-design sessions through innovative approaches, such as Public Participation Geographic Information System (PP-GIS). Site surveys and collaborative sessions involve local administration representatives, residents, neighborhoods, and associations. (See Additional Resources, Box 5 and Table 4, for detailed steps, methods and activities of the UDCW's "Stakeholder Priorities" phase.)

CASE STUDY 8 GOWANUS, NEW YORK CITY URBAN DESIGN CLIMATE WORKSHOP[33]

A 2019 UDCW for the Gowanus canal area of Brooklyn, New York City, focused on urban heat stress adaptation integrated with flood resiliency and GHG mitigation. The Gowanus neighborhood is rapidly being transformed from an industrial area fraught with nineteenth- and twentieth-century pollution issues into a twenty-first-century residential and commercial community. The Urban Land Institute's Gowanus Technical Advisory Panel acknowledged that the anticipated Gowanus rezoning

---

[33] See extended version of case study at https://uccrn.ei.columbia.edu/case-studies.

CASE STUDY 8 (cont.)

would likely create greater density in the neighborhood, particularly for residential uses. Unfortunately, more density increases exposure to higher heat and flood risk.

The UDCW Team configured prototype interventions for strategic sites in Gowanus through baseline (business-as-usual) and best practices (climate-aligned urban design). These provide compelling evidence to NYC policy-makers on the value proposition (financial/health/public real co-benefits) from the UDCW evidence-based strategies. With support from the Urban Land Institute and the National Science Foundation, the UDCW worked in close collaboration with stakeholders to co-produce scenarios to achieve climate resilience and net-zero carbon emissions by balancing a measured amount of carbon (or CO_2 equivalents) released with an equal amount sequestered or offset to support existing NYC policy benchmarks.

Urban heat deserts throughout the study area were identified – all of which lack vegetative cover. The Panel recommended strategies that increase vegetation and leverage the network of hidden creeks in Gowanus and the prevailing summer winds to create "paths of respite" throughout the study area.

Figure 25 Climate-aligned scenario for Gowanus, Brooklyn, New York City.
(NYIT SoAD Urban Design Climate Lab, 2019)

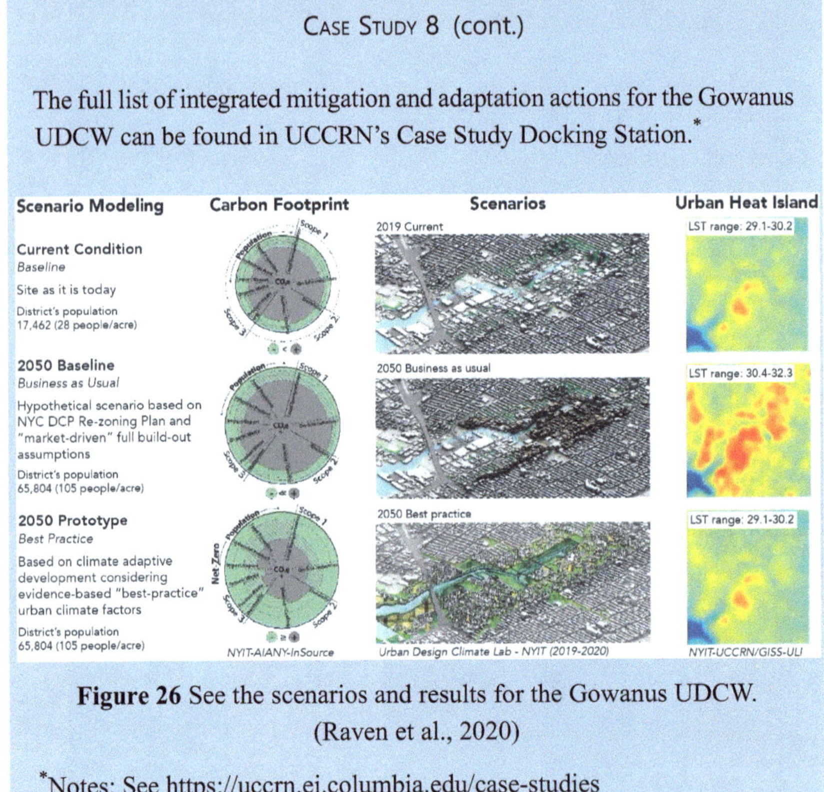

Figure 26 See the scenarios and results for the Gowanus UDCW. (Raven et al., 2020)

*Notes: See https://uccrn.ei.columbia.edu/case-studies

A major goal is to develop a shared understanding of the urban system network in relation to the environmental, functional, spatial, and socioeconomic contexts. Aspirational relationships among stakeholder groups are explored as a component of the future scenarios. (For further discussion of integrating societal challenges into the fabric of climate-driven planning and design processes, see Section 6; See Section 7 for an explanation of how community experts, policymakers, and practitioners strengthen knowledge sharing, co-design, and co-evaluation.)

9.4 Co-Design

Urban transformation pathways are shaped by demographic projections, socio-economic considerations, and driven by stakeholder aspirations or policy mandates (see Section 5). The co-design process demonstrates how urban design, planning and development confront the nexus between climate hazards, urban systems, and low-carbon development. Urban Design Climate Workshop participants draw from a wide range of cross-sectoral stakeholders. The process challenges contemporary rules of engagement in shaping the contemporary city.

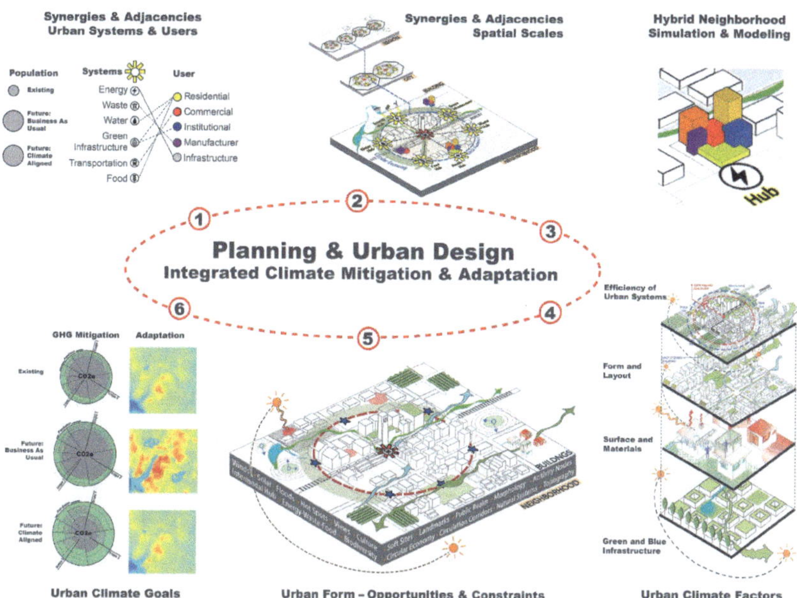

Figure 27 Phasing of the Urban Design Climate Workshop process.
(Raven, 2025)

The UDCWs often occur in cities whose strategic, vulnerable neighborhoods face large-scale rezoning or development pressures. Outputs range from guidelines to analogue diagrams, from off-the-shelf digital parametric three-dimensional modeling tools to GIS mapping (Figure 27).

9.5 Outcomes and Feedback

A critical review of outputs from previous UDCW phases focuses on synergies, trade-offs, constraints, and opportunities across urban systems, spatial scales, and stakeholders. Outputs include:

(1) Prototypes identifying functions, built environment and natural systems
(2) Portfolio of meta-design solutions whose replicable approaches are tailored to local conditions
(3) Design prototypes that consider architectural/technical standards and performance benchmarks
(4) Future scenario comparisons enabled by simulation tools and complemented by multicriteria and cost-benefit analyses.

(See Additional Resources, Table 5, for a complete list of UDCW output.)

Points for further research and consideration that have emerged from the UDCW activities include:

- Opportunities for regulatory and urban systems' synergy overlays at the neighborhood scale
- Prioritizing policies integrating climate adaptation (urban heat/flooding) and climate mitigation (net-zero GHG) in the context of competing policy and investment alternatives
- Private-sector investment and actions through policy, metrics, and incentives
- Framing the nexus of climate justice through the lens of health, equity, poverty, race, education, training, and professional qualification opportunities
- Balancing stakeholder priorities with emissions, adaptation, energy, and environmental performance
- In situ or crowd-sourced monitoring information to compare actual project co-benefits of adaptation and mitigation measures with design scenarios
- Follow-up stakeholder workshops as feedback loop opportunities for continuation of the UDCW process.

A common take away from the different UDCWs is that despite the climate action plan and urban resilience strategies that are in place in many cities, a prevalence of "business as usual" approaches in urban (re)development projects is observed, with major risks for historic residents, vulnerable groups, and carbon neutrality goals.

UDCWs represent auxiliary activities with respect to the complex governance, planning and design processes of the cities in which the workshops took place UDCWs help to connect diverse stakeholder visions through evidence-based data, knowledge sharing, creativity and thus contribute to city climate action planning and design.

10 Conclusions and Research Gaps

Configuring cities to confront climate change is a complex task involving the evolution of energy, food, water, transportation, housing, and waste disposal systems that are major drivers of carbon emissions and vulnerable to growing climate impacts. A city is further shaped by human needs for social connectedness, the expression of cultural identity, and connection to nature. Social, cultural, and behavioral patterns are embedded in the structures of urban life, giving rise to lifestyles and consumption patterns unique to each city.

Metropolitan regions, cities, neighborhoods, and buildings are the scales for engaging, testing, applying and replicating climate-aligned strategies. Urban planning, design, and architecture are ideal platforms for this experimentation, empowering cities to adopt innovative tools and practices, driving equitable climate action, driving the change as a significant force for global transformation.

Accelerating climate solutions in cities requires a new collaborative model for multiscale planning and design that integrates research and practice. Urban planners, designers, and architects can move cities forward by:

- Delivering innovative concepts that promote the rapid deployment of urban solutions concurrently supporting carbon neutrality and multi-hazard risk reduction
- Leveraging physical and digital platforms to enhance multi-stakeholder and multi-sectoral collaboration
- Connecting urban climate science to design practice and visioning
- Encouraging investments in appropriate technologies and processes able to accelerate the needed socio-technical-environmental transitions.

Through a transformative system change approach, planning and design practitioners can ensure that solutions are ideated and implemented to address the root causes instead of symptoms of the climate emergency. With this approach, the potential impact of city strategies expands from the narrow and static focus on reducing GHG emissions to delivering on human needs, investigating solutions relevant to the local context beyond a mere "carbon tunnel vision" and adopting a "full iceberg approach" to urban transformation. (See Figure 10 and Figure 3 in Additional Resources.)

Such an approach has the potential to mitigate existing socio-spatial inequalities. Therefore, addressing the needs of vulnerable and marginalized communities is crucial for creating climate-resilient cities. Including studies on local vulnerabilities in planning and design can align urban climate analyses with local priorities, ensuring that mitigation and adaptation strategies are contextually relevant. This helps prevent maladaptation risks, such as green gentrification, where environmental benefits disproportionately favor higher-income groups. Multi-stakeholder, multi-scale, and context-specific urban processes that emphasize justice and equity can improve spatial decision-making, protect rights, and ensure shared responsibilities for climate and community action.

Integrating climatic and policy information across spatial scales is vital for understanding the urban environment and developing effective design and planning responses to climate change. Data-driven approaches are crucial for assessing the value of urban climate strategies. Data collection and computational tools should be accessible to urban practitioners with limited expertise in urban climate science. In this context, machine learning and AI-driven models need to be used with caution, primarily leveraging the current potential of natural language models for "translating" science and data to decision-makers and communities, embracing complexity rather than fostering simplified automated assumptions.

10.1 Research Gaps

10.1.1 Lack of Mitigation and Adaptation Synergies and Trade-Offs

There is limited research that has systematically evaluated mitigation and adaptation synergies and trade-offs for specific contexts, with only sparse published research analyzing methods for their integration (IPCC, 2022). Identified priorities for research include improving understandings of benefits, costs, synergies, trade-offs, and limitations of major mitigation and adaptation options (Field et al., 2014). According to Shrestha and Dhakal (2019), few studies have identified the potential for synergies in energy, infrastructure planning, construction, and transportation sectors. Similarly, research and evidence on context-specific drivers of co-benefits is limited (Gouldson et al., 2018). In particular, there is a lack of co-benefits research in Oceania and Africa, which have very different development needs than those of Western regions (Deng et al., 2017).

10.1.2 Adaptation Planning Lagging behind Mitigation in Cities

While a general gap is observed between climate change policies and their implementation, implementation of adaptation planning is lagging behind mitigation action in urban areas, and that overall, adaptation planning is occurring in a fragmented, and incremental way, often focused on specific sectors with limited consultation. Coherence of approaches and assessment metrics about carbon neutrality, adaptation and spatial/environmental justice goals across governance and planning/design scales (from state-level target and policies, to regional, city, neighborhood and building levels, including improved building codes) requires a coordinated effort by institutions, practitioners and communities. In particular, the impact of multi-center metropolitan characteristics on GHG emissions and adaptation should be better accounted for and verified using data from more urban areas.

10.1.3 Limited Cross-Scale Integration of Digital Tools

Proprietary digital tools deployed by designers at the building and city scale could be better integrated across scales, and more transparent and accessible to aid interdisciplinary climate change actions for facilities and urban development management. Research focuses on open source and planner/designer-friendly assessment tools highlighting climate benefits (mitigation and adaptation) and co-benefits (social, economic, and environmental) through designer-friendly digital twins enabling the comparison of alternative proposed scenarios at city, neighborhood, and building scales, as well as at conceptual and early-stage design levels. Research on ongoing drivers towards future "smart city" proposals could go beyond the technological parameters and proprietary big data dashboards to facilitate a more human-centered and multi-species focus to potentially recognize "more-than-human" stakeholders.

10.1.4 Insufficient Capacity-Building across Populations and Skillsets

Significant research gaps are identified with respect to building capacities across urban groups. Complementary narrative for a more inclusive, diverse, holistic, and multidisciplinary approach to capacity-building – to balance the dominant prevailing narrative which is overly technocratic and thereby "siloed" – should be further developed. This could involve articulating more of a "shared-language" able to cut across disciplinary boundaries, and link multilevel interests in both higher education and professional qualification. This links to the need of reflection on pedagogical approaches and the effectiveness of resources and frameworks developed, with the novel methods aimed at instilling not only knowledge and skills competency, but also a sense of "agency" in students about the procedural mechanisms that hinder or facilitate climate change actions in practice, within and across their daily work as professions.

10.1.5 Integrating Cultural Contexts and Ethical Considerations in Climate-Resilient Development

Further research is needed to address the gaps in knowledge, including the needs of built environment professionals with respect to the consolidating "climate-resilient development" concept and the levels of personal agency that urban professionals have, also reflecting on the adequacy of professional codes of conduct to include climate change as an ethical issue for practitioners. The role of cultural contexts and, where available, Indigenous and/or traditional environmental knowledge, are important to not only tailor and ground professional knowledge and skills competencies to urban community settings and climatic conditions, promoting greater acceptance and adoption, but also to complement the mainstream scientific approaches.

10.2 Way Forward

To scale up and speed up climate action in cities, transformative approaches that integrate novel strategies tackling interconnected technological, environmental, socioeconomic, and governance challenges linked to climate-resilient urban development must be adopted. A shift in mindset is essential, emphasizing equitable, resilient, and net-zero urban transitions that respect biodiversity, protect ecosystems, and promote inclusivity. Achieving these goals requires systemic change across urban scales and systems, fostering innovation and collaboration at every level.

A key priority is to support through innovative planning and design concepts the integration of urban systems across scales, from buildings and neighborhoods to metropolitan regions. Informing projects and plans through advanced risk assessments can help to update building codes, land-use planning, and infrastructure development, ensuring that climate considerations become central to urban decision-making. Housing and infrastructure systems, including water supply, drainage, sanitation, transport, energy, and waste management, play a critical role in enhancing resilience, reducing greenhouse gas emissions, and improving well-being and equity. Adopting holistic frameworks, such as those reflecting on water–energy–food and climate–pollution–biodiversity nexuses, can inform transformative solutions that are both integrative and scalable.

Monitoring and learning are vital components of this transition. Co-creating city-specific indicators, metrics, and environmental sensing technologies to assess responsiveness of urban transformation to community and climate priorities can enable the implementation of climate-resilient urban development pathways towards shared future visions. Interoperable, user-friendly data systems will facilitate the monitoring of diverse policy domains while supporting the needs of local contexts, including informal settlements and marginalized communities.

Neighborhood-scale climate action offers a valuable testing ground for innovative solutions. Effective climate-resilient development at this scale requires engaging a diverse range of stakeholders and addressing local power dynamics. By aligning community priorities with climate science, cities can create actionable plans that meet the specific needs of local authorities, residents, and private actors. These localized efforts can aim for rapid transferability, enabling the adoption of successful approaches across different contexts and scales.

Ultimately, the future of cities lies in the ability of planners, urban designers, architects, engineers, policymakers, developers, and researchers to foster multisectoral collaboration, harnessing innovation, equity, and cutting-edge technology to transform cities into resilient, sustainable hubs to confront not only the climate crisis but to ensure the well-being of people and nature.

Appendix: UCCRN ARC3.3 Stakeholder Soundings

The UDCWs have been used as research-practice cross fertilization platforms and as an opportunity to engage with city stakeholders during the writing process of this ARC3.3 Element.

In particular, the Erasmus+ UCCRN_edu project in collaboration with the National Science Foundation (NSF) featured several UDCWs in the period 2022–2024 to activate dynamic partnerships between researchers, urban policymakers, and local representatives in Europe, Africa, North America, and Latin America.

References

Acuto, M. (2016). Give cities a seat at the top table. *Nature*, *537*(7622), 611–613. https://doi.org/10.1038/537611a.

Adame, F. (2021). Meaningful collaborations can end 'helicopter research.' Nature, d41586-021-01795–1. https://doi.org/10.1038/d41586-021-01795-1

Addor, F. (2021). Extensão tecnológica e Tecnologia Social: reflexões em tempos de pandemia. *NAU Social*, *11*(21), 395–412.

Ajibade, I. (2019). Planned retreat in Global South megacities: disentangling policy, practice, and environmental justice. *Climatic Change*, *157*, 299–317.

Ajuntament de Barcelona. (2022). Refugios climáticos en las Escuelas. https://www.barcelona.cat/barcelona-pel-clima/es/escuelas-refugios-climaticos

Amin, A. (2006). The good city. *Urban Studies*, *43*(5–6), 1009–1023.

Amorim-Maia, A. T., Anguelovski, I., Chu, E., & Connolly, J. (2022). Intersectional climate justice: a conceptual pathway for bridging adaptation planning, transformative action, and social equity. *Urban Climate*, *41*, 101053.

Anand, V., Kadiri, V. L., & Putcha, C. (2023). Passive buildings: a state-of-the-art review. *Journal of Infrastructure Preservation and Resilience*, 4:3, 1–24. https://doi.org/10.1186/s43065-022-00068-z.

Anguelovski, I., Connolly, J., Masip, L., & Pearsall, H. (2018a). Assessing environmental gentrification impacts of neighborhood greening in historically disenfranchised areas: a longitudinal and spatial analysis through the municipal area of Barcelona. *Urban Geography*, *38*(3), 458–491.

Anguelovski, I., Argüelles, L., Baró, F., Cole, H. V. S., Connolly, J. J. T., García Lamarca, M., Loveless, S., Pérez del Pulgar, C., Shokry, G., Trebic, T., & Wood, E. (2018b). *Green Trajectories: Municipal Policy Trends and Strategies for Greening in Europe*. Barcelona Lab for Urban Environmental Justice and Sustainability. https://ddd.uab.cat.

Anguelovski, I., Shi, L., Chu, E., Gallagher, D., Goh, K., Lamb, Z., Reeve, K., & Teicher, H. (2016). Equity impacts of urban land use planning for climate adaptation: critical perspectives from the Global North and South. *Journal of Planning Education and Research*, *36*(3), 333–348.

Archer, D., Almansi, F., DiGregorio, M., Roberts, D., Sharma, D., & Syam, D. (2014). Moving towards inclusive urban adaptation: approaches to integrating community based adaptation to climate change at city and national scale. *Climate and Development*, *6*(4), 345–356.

Arnett, H. (2023). Professional Agency for Urban Health Design: A Call for Collective Responsibility. Healthy City Design International 2022 review. *Cities & Health*, 8(5), 850–853.

Arteaga, S. (2023a). *Community Mapping and Participatory Planning Workshop during the Development of the Apartadó River Master Plan* [Photograph]. Personal collection.

Arteaga, S. (2023b). *The Apartadó River Crosses the City from East to West; Units Are Expressed in Meters* [Photograph]. Personal collection.

Asadzadeh, A., Fekete, A., Khazai, B., Moghadas, M., Zebardast, E., Basirat, M., & Kötter, T. (2023). Capacitating urban governance and planning systems to drive transformative resilience. *Sustainable Cities and Society, 96*, 104637.

Avila-Palencia, I., & Lutzu, J. (2023). *Superblocks in Horta and Sant Antoni* [Photograph]. Personal collection.

Ayers, J., & Forsyth, T. (2009). Community-based adaptation to climate change: strengthening resilience through development. *Environment, 51*(4), 22–31.

Aylett, A. (2015). Institutionalizing the urban governance of climate change adaptation: results of an international survey. *Urban Climate, 14*, 4–16. https://doi.org/10.1016/j.uclim.2015.06.005.

Babagoli, M. A., Kaufman, T. K., Noyes, P., & Sheffield, P. E. (2019). Exploring the health and spatial equity implications of the New York City Bike share system. *Journal of Transport & Health, 13*, 200–209.

Bachra, S., Lovell, A., McLachlan, C., & Minas, A. M. (2020). *The Co-Benefits of Climate Action: Accelerating City-Level Ambition.* London: CDP Worldwide. https://cdn.cdp.net.

Bain, P. G., Milfont, T. L., Kashima, Y., et al. (2016). Co-benefits of addressing climate change can motivate action around the world. *Nature Climate Change, 6*(2), 154–157.

Barau, A. S. (2020). From zero-acreage farming to zero hunger in African cities: some possibilities and opportunities. In V. Squires & M. Gaur (Eds.), *Food Security and Land Use Change under Conditions of Climatic Variability.* Cham: Springer. https://doi.org/10.1007/978-3-030-36762-6_11.

Béné, C., Mehta, L., McGranahan, G., Cannon, T., Gupte, J., & Tanner, T. (2018). Resilience as a policy narrative: potentials and limits in the context of urban planning. *Climate and Development, 10*(2), 116–133.

Berardi, U., Jandaghian, Z., & Graham, J. (2020). Effects of greenery enhancements for the resilience to heat waves: a comparison of analysis performed through mesoscale (WRF) and microscale (Envi-met) modeling. *Science of The Total Environment, 747*, 141300.

Biggs, J., & Tang, C. (2007). *Teaching for Quality Learning at University* (3rd ed.). Milton Keynes: Open University Press.

Biljecki, F., & Ito, K. (2021). Street view imagery in urban analytics and GIS: a review. *Landscape and Urban Planning, 215*, 104217.

Binarti, F., Koerniawan, M. D., Triyadi, S., Utami, S. S., & Matzarakis, A. (2020). A review of outdoor thermal comfort indices and neutral ranges for hot-humid regions. *Urban Climate, 31, 100531*.

Birchall, S. J., & Bonnett, N. (2021). Climate change adaptation policy and practice: the role of agents, institutions and systems. *Cities, 108*, 103001.

Bouyé, M., D. O'Connor, A. Tankou, D. Grinspan, D. Waskow, S. Chattopadhyay, and A. Scott. 2020. "Achieving Social Equity in Climate Action: Untapped Opportunities and Building Blocks for Leaving No One Behind." Working Paper. Washington, DC: World Resources Institute. https://doi.org/10.46830/wriwp.19.00090.

Bremer, S., Wardekker, A., Dessai, S., Sobolowski, S., Slaattelid, R., & van der Sluijs, J. (2019). Toward a multi-faceted conception of co-production of climate services. *Climate Services, 13*, 42–50.

British Columbia Society of Landscape Architects. (2020). Integrated Building Adaptation and Mitigation Assessment (IBAMA). https://www.bcsla.org/content/integrated-building-adaptation-and-mitigation-assessment-ibama

Bulkeley, H., & Betsill, M. M. (2013). Revisiting the urban politics of climate change. Environmental politics, 22(1), 136–154.

Bulkeley, H., Edwards, G. A. S., & Fuller, S. (2014). Contesting climate justice in the city: examining politics and practice in urban climate change experiments. *Global Environmental Change, 25*, 31–40.

Bush, J., & Doyon, A. (2019). Building urban resilience with nature-based solutions: how can urban planning contribute? *Cities, 95*, 102483.

C40 Cities. (2019). *Inclusive Planning: Executive Guide.* www.c40knowledgehub.org/s/article/Inclusive-Planning-Executive-Guide?language=en_US.

C40 Cities Climate Leadership Group. (2018). Adaptation and Mitigation Interaction Assessment (AMIA) tool. www.c40knowledgehub.org/s/article/Adaptation-and-Mitigation-InteractionAssessment-AMIA-tool?language=en_US.

C40 Cities Climate Leadership Group. (2024). *Climate Action Guide for Urban Planners.* www.c40knowledgehub.org/s/article/Climate-action-guide-for-urban-planners?language=en_US.

C40 Knowledge Hub. (2018). Adaptation and Mitigation Interaction Assessment (AMIA) tool. https://www.c40knowledgehub.org/s/article/Adaptation-and-Mitigation-Interaction-Assessment-AMIA-tool?language=en_US.

Cabeza, L. F., Bai, Q., Bertoldi, P., Kihila, J. M., Lucena, A. F. P., Mata, É., Mirasgedis, S., Novikova, A., & Saheb, Y. (2022). Buildings. In P. R. Shukla, J. Skea, R. Slade, A. Al Khourdajie, R van Diemen, D. McCollum, M.

Pathak, S. Some, P. Vyas, R. Fradera, M. Belkacemi, A. Hasija, G. Lisboa, S. Luz, & J. Malley (Eds.), *Climate Change 2022: Mitigation of Climate Change. Contribution of Working Group III to the Sixth Assessment Report of the Intergovernmental Panel on Climate Change.* Cambridge: Cambridge University Press.

Calzada, I., Pérez-Batlle, M., & Batlle-Montserrat, J. (2021). People-centered smart cities: an exploratory action research on the cities' coalition for digital rights. *Journal of Urban Affairs 45(9), 1537–1562. https://doi.org/10.1080/0735 2166.2021.1994861*

Canadian Institute of Planners. (2019). *Canadian Institute of Planners 2019 Climate Change Survey.* www.cip-icu.ca.

Cansino, J. M., Román, R., & Rueda-Cantuche, J. M. (2015). Will China comply with its 2020 carbon intensity commitment? *Environmental Science & Policy, 47,* 108–117. https://doi.org/10.1016/j.envsci.2014.11.004.

CARE. (2010). *Toolkit for Integrating Climate Change Adaptation into Development Projects.* https://careclimatechange.org/wp-content/uploads/ 2019/06/CARE_Integration_Toolkit.pdf.

Castan Broto, V. (2021). *Just Climate Adaptation in Cities: Reflections for an Interdisciplinary Research Agenda.* The British Academy.

Chang, S. E., Yip, J. Z. K., & Tse, W. (2018). Effects of urban development on future multi-hazard risk: the case of Vancouver, Canada. *Natural Hazards,* 98, 251–265.

Chen, F., Kusaka, H., Bornstein, R., Ching, J., Grimmond, C. S. B., Grossman-Clarke, S., Loridan, T., Manning, K. W., Martilli, A., Miao, S., & Sailor, D. (2011). The integrated WRF/urban modelling system: development, evaluation, and applications to urban environmental problems. *International Journal of Climatology,* 31(2), 273–288.

Chen, I. H. (2021). New conceptual framework for flood risk assessment in Sheffield, UK. *Geographical Research.*

Chen, H. (2023). *This Country's Love Affair With Air Conditioning Shows a Catch 22 of Climate Change.* CNN. www.cnn.com/2023/06/09/asia/air-conditioning-singapore-climate-change-intl-hnk-dst/index.html#:~:text=Indeed%2C%20in %20this%20city%2C%20air,as%20they%20cross%20a%20street.

Chester, M. V., Miller, T. R., Muñoz-Erickson, T. A., Helmrich, A. M., Iwaniec, D. M., McPhearson, T., Cook, E. M., Grimm, N. B., & Markolf, S. A. (2023). Sensemaking for entangled urban social, ecological, and technological systems in the Anthropocene. *NPJ Urban Sustainability, 3* (1), 39. https://doi.org/10.1038/s42949-023-00120-1.

Chow, W. T. L., & Roth, M. (2006). Temporal dynamics of the urban heat island of Singapore. *International Journal of Climatology, 26*(15), 2243–2260. https://doi.org/10.1002/joc.1364.

Chow, W. T. L., Brennan, D., & Brazel, A. J. (2012). Urban heat island research in Phoenix, Arizona: theoretical contributions and policy applications. *Bulletin of the American Meteorological Society*, *93*(4), 517–530. https://doi.org/10.1175/BAMS-D-11-00011.1.

Christchurch City Council. (2021). *Ōtautahi Christchurch Climate Resilience Strategy*. https://ccc.govt.nz/assets/Documents/Environment/Climate-Change/Otautahi-Christchurch-Climate-Resilience-Strategy.pdf.

Christiansen, L., Martinez, G., & Naswa, P. (Eds.). (2018). *Development of Universal Metrics for Adaptation Effectiveness. Adaptation Metrics: Perspectives on measuring, aggregating and Comparing Adaptation Results*. Copenhagen: UNEP DTU Partnership.

Chu, E., Anguelovski, I., & Carmin, J. (2016). Inclusive approaches to urban climate adaptation planning and implementation in the Global South. *Climate Policy*, 16(3), 372–392.

City of Barcelona. (2018). *Climate Plan 2018–2030*. https://cdn.locomotive.works/sites/5ab410c8a2f42204838f797e/content_entry5ae2f905a2f4220ae645f026/5afc10d27478206be9209e60/files/Bcn_Climate_Plan.pdf?1526468818

City of Barcelona. (2020). *Towards Superblock Barcelona*.

City of Berlin. (2016). *Climate- Neutral Berlin 2050*. www.berlin.de/sen/uvk/_assets/klimaschutz/publikationen/machbarkeitsstudie_berlin2050_en.pdf.

City of Berlin. (2018). *Berlin Energy and Climate Protection Programme 2030 (BEK)*. www.berlin.de/sen/uvk/_assets/klimaschutz/publikationen/bek2030_broschuere_en.pdf.

City of Hong Kong. (2021). *Clean Air Plan for Hong Kong 2035*. https://www.eeb.gov.hk/sites/default/files/pdf/Clean_Air_Plan_2035_eng.pdf

City of Hong Kong. (2021). *Climate Action Plan 2050*. https://cnsd.gov.hk/wp-content/uploads/pdf/CAP2050_leaflet_en.pdf

City of Hong Kong. (2020). *Green Tech Fund*. https://www.gtf.gov.hk/en/index.html

City of Melbourne. (2017). *Climate Change Adaptation Strategy*. https://www.melbourne.vic.gov.au/climate-change-adaptation-strategy

City of Melbourne. (2019). *Future Melbourne Committee Minutes*. https://mvga-prod-files.s3.ap-southeast-4.amazonaws.com/public/about-council/committees-meetings/meeting-archive/MeetingAgendaItemAttachments/864/JUL19%2520FMC2%2520CONFIRMED%2520MINUTES.PDF

City of Melbourne. (2021). *Emissions Reduction Plan*. https://www.melbourne.vic.gov.au/emissions-reduction-plan

City of Vancouver. (2025). *Climate Emergency Action Plan*. https://vancouver.ca/green-vancouver/climate-emergency-action-plan-in-depth.aspx

City of Melbourne. (2025). *Melbourne Planning Scheme.* https://planning-schemes.app.planning.vic.gov.au/Melbourne/ordinance

City of Tokyo. (2011). *Tokyo Climate Change Strategy.* www.kankyo.metro.tokyo.lg.jp/documents/d/kankyo/outline-of-zero-emission-tokyo-strategy.

City of Vancouver. (2016). *Zero Emissions Building Plan.* https://vancouver.ca/files/cov/zero-emissions-building-plan.pdf

City of Vancouver. (2019). Rain City Strategy: A green rainwater infrastructure and rainwater management initiative. https://vancouver.ca/files/cov/rain-city-strategy.pdf

City of Vancouver. (2025). *Climate Change Adaptation Strategy.* https://vancouver.ca/green-vancouver/climate-change-adaptation-strategy.aspx

City of Vancouver. (2025). *Climate Emergency Action Plan.* https://vancouver.ca/green-vancouver/climate-emergency-action-plan-in-depth.aspx

Clarke, P. (2017). *The Façade of Nightingale 1* [Photograph]. @peterclarke-photo. Latitude.

Climate-ADAPT. (2022). *Paris Oasis Schoolyard Programme, France.* https://climate-adapt.eea.europa.eu/en/metadata/case-studies/paris-oasis-school yard-programme-france

Coalition for Urban Transitions. (2019). *Climate Emergency, Urban Opportunity.* https://urbantransitions.global/en/publication/climate-emer gency-urban-opportunity/.

Cohen, D. (2018). *Climate Justice and the Right to the City.* Penn: Current Research On Sustainable Urban Development, Penn Institute for Urban Research.

Commonwealth Association of Architects, Commonwealth Association of Planners, Commonwealth Association of Surveying and Land Economy, Commonwealth Engineers Council. (2020). *Survey of the Built Environment Professions, 2020.* https://commonwealthsustainablecities.org/wp-content/uploads/2020/06/200621-Commonwealth-Sustainable-Urbanisation-Programme-Launch-of-the-Survey-of-the-Profession.pdf.

Connolly, J. J. T., & Anguelovski, I. (2021). Three Histories of Greening and Whiteness in American Cities. *Frontiers in Ecology and Evolution, 9,* 621783. https://doi.org/10.3389/fevo.2021.621783.

Cooperative for Assistance and Relief Everywhere (CARE). (2010). *Community-Based Adaptation Toolkit.*

Corning, P. A. (1998). "The synergism hypothesis": on the concept of synergy and its role in the evolution of complex systems. *Journal of Social and Evolutionary Systems, 21*(2), 133–172.

Costello, A., & Zumla, A. (2000) Moving research partnerships in developing countries. *BMJ,* 321, 827–829.

Crichton, D. (1999). The risk triangle. In Ingleton, J., Ed., *Natural disaster management*, Tudor Rose, London, 102–3.

Dagnino, R. (2010). *Tecnologia social: ferramenta para construir outra sociedade* (2nd ed.). Campinas: Komedi.

Dagnino, R. (2014). *Tecnologia Social: contribuições conceituais e metodológicas*. Campina Grande: EDUEPB.

Davey, T. (2025). The spatial politics of an urban carbon accounting standard. Environment and Planning E: Nature and Space, 25148486251315290. https://doi.org/10.1177/25148486251315290

De Abreu, V. H. S., Santos, A. S., & Monteiro, T. G. M. (2022). Climate change impacts on the road transport infrastructure: a systematic review on adaptation measures. *Sustainability, 14*(14), 8864. https://doi.org/10.3390/su14148864.

de Jong, M., Yu, C., Joss, S., Wennersten, R., Yu, L., Zhang, X., & Ma, X. (2016). Eco city development in China: addressing the policy implementation challenge. *Journal of Cleaner Production, 134*, 31–41.

Demuzere, M., Orru, K., Heidrich, O., Olazabal, E., Geneletti, D., Orru, H., … Faehnle, M. (2014). Mitigating and adapting to climate change: multi-functional and multi-scale assessment of green urban infrastructure. *Journal of Environmental Management, 146*, 107–115.

Deng, H. M., Liang, Q. M., Liu, L. J., & Anadon, L. D. (2017). Co-benefits of greenhouse gas mitigation: a review and classification by type, mitigation sector, and geography. *Environmental Research Letters, 12*(12), 123001.

Deutscher Wetterdienst. (2024). *The German Weather Service (DWD) Berlin.* www.dwd.de/EN/Home/home_node.html.

Diezmartínez, C. V., & Short Gianotti, A. G. (2022). US cities increasingly integrate justice into climate planning and create policy tools for climate justice. *Nature Communications, 13*(1), 5763.

Dodman, D., Hayward, B., Pelling, M., Castan Broto, V., Chow, W., Chu, E., Dawson, R., Khirfan, L., McPhearson, T., Prakash, A., Zheng, Y., & Ziervogel, G. (2022). Cities, Settlements and Key Infrastructure. In: *Climate Change 2022: Impacts, Adaptation and Vulnerability. Contribution of Working Group II to the Sixth Assessment Report of the Intergovernmental Panel on Climate Change* H.-O. Pörtner, D. C. Roberts, M. Tignor, E. S. Poloczanska, K. Mintenbeck, A. Alegría, M. Craig, S. Langsdorf, S. Löschke, V. Möller, A. Okem, B. Rama (eds.)]. Cambridge: Cambridge University Press, 907–1040, doi:10.1017/9781009325844.008.

Douglas, E. M., Reardon, K. M., & Täger, M. C. (2018). Participatory action research as a means of achieving ecological wisdom within climate change resiliency planning. *Journal of Urban Management, 7*(3), 152–160.

Duarte, F., & deSouza, P. (2020). Data science and cities: a critical approach. *Harvard Data Science Review.*

Elmqvist, T., Andersson, E., Frantzeskaki, N., McPhearson, T., Olsson, P., Gaffney, O., ... & Folke, C. (2019). Sustainability and resilience for transformation in the urban century. *Nature Sustainability, 2*(4), 267–273.

Endo, I., Magcale-Macandog, D. B., Kojima, S., Johnson, B. A., Bragais, M. A., Macandog, P. B. M., & Scheyvens, H. (2017). Participatory land-use approach for integrating climate change adaptation and mitigation into basin-scale local planning. *Sustainable Cities and Society, 35*, 47–56.

Ensor, J., & Berger, R. P. (2009). *Community-based adaptation and culture in theory and practice.* In N. Adger, I. Lorenzoni, and K. L. O'Brien (Eds.), *Adapting to Climate Change* (pp. 227–239). Cambridge: Cambridge University Press. https://doi.org/10.1017/cbo9780511596667.015.

European Commission. (2008). *The European Qualifications Framework for Lifelong Learning.* Luxembourg, Office for Official Publications of the European Communities. http://ecompetences.eu.

European Commission. (2021). *An Ecological Corridor to Reunite the City and Nature. CORDIS – EU Research Results.* https://cordis.europa.eu/article/id/428824-an-ecological-corridor-to-reunite-the-city-and-nature.

European Commission. (2023). *Employment and social developments in Europe 2023.* Publications Office of the European Union. Luxembourg.

Fallmann, J., & S. Emeis. (2020). How to bring urban and global climate studies together with urban planning and architecture? *Developments in the Built Environment,* 4, 100023.

FEMA. (2010). Natural Hazards and Sustainability for Residential Buildings. https://biotech.law.lsu.edu/blaw/FEMA/FEMA-Sustainable-Design_092210.pdf

Ficklin, L., Stringer, L. C., Dougill, A. J., & Sallu, S. M. (2017). Climate compatible development reconsidered: calling for a critical perspective. *Climate and Development, 10*(3), 193–196.

Field, C. B., Barros, V. R., Dokken, D. J., Mach, K. J., & Mastrandrea, M. D. (Eds.). (2014). Climate-resilient pathways: adaptation, mitigation, and sustainable development. In *Climate Change 2014 Impacts, Adaptation, and Vulnerability* (pp. 1101–1131). Cambridge: Cambridge University Press. https://doi.org/10.1017/CBO9781107415379.025.

Fitzgibbons, J., & Mitchell, C. (2019). Just urban futures? Exploring equity in "100 Resilient Cities." *World Dev.* 122, 648–659.

Floater, G., Heeckt, C., Ulterino, M., Mackie, L., Rode, P., Bhardwaj, A., Carvalho, M., Gill, D., Bailey, T., & Huxley, R. (2016). *Co-Benefits of*

Urban Climate Action: A Framework for Cities. London School of Economics and Political Science.

Fong, W. K., Sotos, M., Doust, M., Schultz, S., Marques, A., & Deng-Beck, C. (2021). *Global Protocol for Community-Scale Greenhouse Gas Inventories*. www.ghgprotocol.org.

Forsyth, T. (2013). Community-based adaptation: a review of past and future challenges – community-based adaptation. *Wiley Interdisciplinary Reviews: Climate Change, 4*(5), 439–446.

Foster, S., Leichenko, R., Nguyen, K. H., Blake, R., Kunreuther, H., Madajewicz, M., Petkova, E. P., Zimmerman, R., Corbin-Mark, C., Yeampierre, E., Tovar, A., Herrera, C., & Ravenborg, D. (2019). New York City panel on climate change 2019 report chapter 6: community-based assessments of adaptation and equity. *Annals of the New York Academy of Sciences*, 1439(1), 126–173.

Fox, A., Ziervogel, G., & Scheba, S. (2021). *Strengthening Community-Based Adaptation for Urban Transformation: Managing Flood Risk in Informal Settlements in Cape Town*. ACDI. https://acdi.uct.ac.za/sites/default/files/content_migration/acdi_uct_ac_za/1205/files/Fox%2520et%2520al%252C%2520Strengthening%2520CBA%2520for%2520%2520urban%2520transformation_0.pdf.

Friedmann, J. (2000). The good city: in defense of utopian thinking. *International Journal of Urban and Regional Research*, 24(2), 460–472.

Friesenecker, M., Thaler, T., & Clar, C. (2024). Green gentrification and changing planning policies in Vienna? *Urban Research & Practice, 17*(3), 393–415. https://doi.org/10.1080/17535069.2023.2228275.

Fünfgeld, H., & Schmid, B. (2020). Justice in climate change adaptation planning: conceptual perspectives on emergent praxis. *Geographica Helvetica, 75*(4), 437–449.

Future Earth. (2017). *Spotlight on SDG Labs: Trees grow in Kano*. Millennials and Resilience: City, Innovation and Transformation of Youths Laboratory. https://futureearth.org/2017/08/23/spotlight-on-sdg-labs-trees-grow-in-kano/.

Galende-Sánchez, E. and Sorman, A. H. (2021). From consultation toward co-production in science and policy: a critical systematic review of participatory climate and energy initiatives. *Energy Research & Social Science, 73*, 101907. https://doi.org/10.1016/j.erss.2020.101907.

Gaziulusoy, İ., & Ryan, C. (2017) Shifting conversations for sustainability transitions using participatory design visioning, *The Design Journal*, 20: sup1, S1916–S1926.

GCoM. (2022). *Energizing City Climate Action: The 2022 Global Covenant of Mayors Impact Report*. www.globalcovenantofmayors.org.

GEMET. (2021). Ecological capacity definition. https://www.eionet.europa.eu/gemet/en/concept/15125

GlobalABC /IEA/UNEP. (2020). *GlobalABC Roadmap for Buildings and Construction: Towards a Zero-Emission, Efficient and Resilient Buildings and Construction Sector*. Paris: IEA.

Gobierno de Mérida. (2018). *Actualización del plan de acción climática municipal*. www.merida.gob.mx/sustentable/contenidos/doc/PACMUN_21JUL21.pdf.

Gondhalekar, D., Bhaduri, A., Waibel, D., & Qureshi, A. (2023, May 15). *Assessment of centralized and decentralized water reclamation with resource recovery in Leh, India by taking a Nexus approach and applying a MCDA framework*. https://doi.org/10.5194/egusphere-egu23-9676.

Gonzalez, R., James, T., & Ross, J. (2017). *Community-driven climate resilience planning: a framework*. National Association of Climate Resilience Planners. https://movementstrategy.org/wp-content/uploads/2021/10/Community-Driven-Climate-Resilience-Planning-A-Framework.pdf.

Gouldson, A., Sudmant, A., Khreis, H., & Papargyropoulou, E. (2018). *The Economic and Social Benefits of Low-Carbon Cities: A Systematic Review of the Evidence*. Coalition for Urban Transitions. http://newclimateeconomy.net/content/cities-working-papers.

Grafakos, S., Pacteau, C., Delgado, M., Landauer, M., Lucon, O., & Driscoll, P. (2018). Integrating mitigation and adaptation: opportunities and challenges. In Rosenzweig, C., Solecki, W., Romero-Lankao, P., Mehrotra, S., Dhakal, S., & Ali Ibrahim, S. (Eds.), *Climate Change and Cities: Second Assessment Report of the Urban Climate Change Research Network*. New York: Cambridge University Press, 101–138.

Grafakos, S., Trigg, K., Landauer, M., Chelleri, L., & Dhakal, S. (2019). Analytical framework to evaluate the level of integration of climate adaptation and mitigation in cities. *Climatic Change, 154*(1–2), 87–106.

Grainger, G. (2022). *To Create Net-Zero Cities, We Need to Look Hard at Our Older Buildings*. World Economic Forum. www.weforum.org/stories/2022/11/net-zero-cities-retrofit-older-buildings-cop27/.

Grainger, G. (2024, January 22). *2024: A Tipping Point for Investing in Sustainable Buildings*. World Economic Forum. www.weforum.org/stories/2024/01/sustainable-office-buildings/.

Graves, R. (2020). The Paradox of Metrics: Setting Goals for Regenerative Design and Development. *Ecologies design: Transforming Architecture, Landscape, and Urbanism*, 34–43.

GRESB. (2023). *Real Estate Standard and Reference Guide*. www.documents
.gresb.com.

Gromala, D. S., Kapur, O., Kochkin, V., Line, P., Passman, S., Reeder, A., &
Trusty, W. (2010). *Natural Hazards and Sustainability for Residential
Buildings*. Washington, DC: FEMA.

Habitat for Humanity. (2021). *Housing and the Sustainable Development
Goals: The Transformational Impact of Housing*. www.habitat.org/sites/
default/files/documents/Housing-and-Sustainable-Development-Goals.pdf.

Hagan, S. (2022). *Revolution? Architecture and the Anthropocene*. Lund
Humphries Publishers.

Hamin, E. M., & Gurran, N. (2009). Urban form and climate change: balancing
adaptation and mitigation in the US and Australia. *Habitat International*, *33*
(3), 238–245.

Han, H., Bai, X., & Dong, L. (2025). Global policy stocktake of urban climate
resilience: A literature review. Resources, Conservation and Recycling, 212,
107923. https://doi.org/10.1016/j.resconrec.2024.107923

Han, J., Mo, N., Cai, J., Ouyang, L., & Liu, Z. (2024). Advancing the local
climate zones framework: a critical review of methodological progress,
persisting challenges, and future research prospects. *Humanities and Social
Sciences Communications*, *11*(1), 538. https://doi.org/10.1057/s41599-024-
03072-8.

Hansen, G. (2012). When students design learning landscapes: designing for
experiential learning through experiential learning. *NACTA Journal*, *56*(4),
30–35.

Hardoy, J., Gencer, E., & Winograd, M. (2019). Participatory planning for
climate resilient and inclusive urban development in Dosquebradas, Santa
Ana and Santa Tomé. *Environment and Urbanization*, *31*(1), 33–52.

Hartenberger, U., Lorenz, D., & Lützkendorf, T. (2013). A shared built environment professional identity through education and training. *Building
Research & Information*, *41*(1), 60–76.

Harvey, D. (1973). *Social Justice and the City*. Blackwell.

Hasell, J., Arriagada, P., Ortiz-Ospina, E., & Roser, M. (2023). *Economic
Inequality*. Our World in Data. https://ourworldindata.org/economic-inequal
ity#explore-data-on-economic-inequality.

Heimann, T., & Mallick, B. (2016). Understanding climate adaptation cultures
in global context: proposal for an explanatory framework. *Climate*, *4*(4), 59.
https://doi.org/10.3390/cli4040059.

Hess, D. J., & McKane, R. G. (2021). "Making Sustainability Plans More
Equitable: An Analysis of 50 US Cities." *Local Environment*, 1–16.

Hoff, H. (2011). Understanding the nexus, background paper for the Bonn 2011 Conference: The water, energy and food security nexus. Stockholm Environment Institute (SEI), Stockholm, Sweden. www.sei.org/mediamana ger/documents/Publications/SEI-Paper-Hoff-UnderstandingTheNexus-2011.pdf

Hölscher, K., Frantzeskaki, N., McPhearson, T., & Loorbach, D. (2019). Capacities for urban transformations governance and the case of New York City. Cities, 94, 186–199.

Hosey, L. (2017). Where Architects Stand on Climate Change. *Architectural Record*, April 11, 2017.

Howard, L. (1818). *The climate of London. Deduced from meteorological observations, made at different places in the neighbourhood of the metropolis.* Published by W. Phillips, George Yard, Lombard Street, London.

Hsu, C. W., & Fingerman, K. (2021). Public electric vehicle charger access disparities across race and income in California. *Transport Policy, 100,* 59–67.

Hua, J., Zhang, X., Ren, C., Shi, Y., & Lee, T.-C. (2021). Spatiotemporal assessment of extreme heat risk for high-density cities: a case study of Hong Kong from 2006 to 2016. *Sustainable Cities and Society, 64,* 102507. https://doi.org/10.1016/j.scs.2020.102507

Hughes, S, & Hoffmann, M. (2020). Just urban transitions: toward a research agenda. *WIREs Clim. Change, 11:e640.*

Hughes, S., Tozer, L., & Giest, S. (2019). The politics of data-driven urban climate change mitigation. In J. van der Heijden, H. Bulkeley, & C. Certomà (Eds.), *Urban Climate Politics: Agency and Empowerment* (pp. 116–134). Cambridge University Press. https://www.cambridge.org/core/books/abs/ urban-climate-politics/politics-of-datadriven-urban-climate-change-mitiga tion/D921A10D12B4F626C881AFA3608FB824

Hürlimann, A. C., Browne, G., Warren-Myers, G., & Francis, V. (2018). Facilitating climate change adaptation in the Australian construction industry-identification of information needs. Proceedings of the 4th Practical Responses to Climate Change Conference:" *Climate Adaptation 2018: Learn, Collaborate, Act*, 155–163.

Hürlimann, A. C., Moosavi, S., & Browne, G. R. (2021a) Climate change transformation: a definition and typology to guide decision making in urban environments. *Sustainable Cities and Society, 70, 102890. https://doi.org/ 10.1016/j.scs.2021.10289070.*

Hürlimann, A. C., Moosavi, S., & Browne, G. R. (2021b). Urban planning policy must do more to integrate climate change adaptation and mitigation actions. *Land Use Policy, 101,* 1–9.

Hürlimann, A., Cobbinah, P. B., Bush, J., & March, A. (2021c). Is climate change in the curriculum? An analysis of Australian urban planning degrees. *Environmental Education Research, 27*(7), 970–991.

Hürlimann, A. C., Warren-Myers, G., Nielsen, J., Moosavi, S., Bush, J., & March, A. (2022a). Towards the transformation of cities: A built environment process map to identify the role of key sectors and actors in producing the built environment across life stages. *Cities, 121*(February), 103454.

Hürlimann, A. C., Nielsen, J., Moosavi, S., Bush, J., Warren-Myers, G., & March, A. (2022b). Climate change preparedness across sectors of the built environment: a review of literature. *Environmental Science & Policy, 128*, 277–289.

Hürlimann, A., Beilin, R., & March, A. (2023a). "Rethinking the way we practice our professions": social-ecological resilience for built environment professionals. *Journal of Further and Higher Education, 47*(1), 118–133.

Hürlimann, A., Cobbinah, P., Bush, J., & Gaisie, E. (2023b). Continuing education for climate change: a study of Australian urban planners' current practices and developing competency. *Journal of Planning Education and Research, 44*(4), 2316–2336. https://doi.org/10.1177/0739456X231171100.

IAA Mobility. (2024). *Superblocks for Everyone!* www.iaa-mobility.com/en/newsroom/news/urban-mobility/superblocks-for-everyoneICLEIUSA. (2019). *US Community Protocol for Accounting and Reporting of Greenhouse Gas Emissions*. www.icleiusa.org.

IEA. (2021). Empowering Cities for a Net Zero Future, IEA, Paris. www.iea.org/reports/empowering-cities-for-a-net-zero-future.

IFLA. (2019, September 16). *IFLA Declares a Climate and Biodiversity Emergency.* www.iflaworld.com/newsblog/ifla-declares-a-climate-and-biodiversity-emergency.

International Finance Corporation. (2019). *Green Buildings: A Financial and Policy Blueprint for Emerging Markets.* www.ifc.org/en/insights-reports/2019/green-buildings-reportwww.ifc.org/en/insights-reports/2019/green buildings-report.

International Passive House Association. (2018). *Active for more comfort: Passive House.* https://passivehouse-international.org/upload/GRBR_EN_2018_Sammelmappe/GRBR_EN_2018_Sammelmappe.html

IPCC. (2014). *Climate Change 2014: Synthesis Report. Contribution of Working Groups I, II and III to the Fifth Assessment Report of the Intergovernmental Panel on Climate Change* [Core Writing Team, R. K. Pachauri and L. A. Meyer (eds.)]. IPCC, Geneva, Switzerland.

IPCC. (2018). *Global Warming of 1.5°C. An IPCC Special Report on the impacts of global warming of 1.5°C above pre-industrial levels and related global greenhouse gas emission pathways, in the context of strengthening the global response to the threat of climate change, sustainable development, and efforts to eradicate poverty.* Cambridge: Cambridge University Press.

IPCC. (2022). *Climate Change 2022: Impacts, Adaptation and Vulnerability. Contribution of Working Group II to the Sixth Assessment Report of the Intergovernmental Panel on Climate Change.* Cambridge University Press. Cambridge: Cambridge University Press.

IPCC. (2023). Summary for Policymakers. In: *Climate Change 2023: Synthesis Report. Contribution of Working Groups I, II and III to the Sixth Assessment Report of the Intergovernmental Panel on Climate Change* [Core Writing Team, H. Lee and J. Romero (eds.)]. IPCC, Geneva, Switzerland, 1–34, doi: 10.59327/IPCC/AR6-9789291691647.001.

Ipsos (2021). UN Sustainable Development Goals in 2021: Public Opinion on Priorities and Stakeholders' Commitment 28-Country Ipsos survey for The World Economic Forum.www.ipsos.com/sites/default/files/ct/news/docu ments/2021-05/April-May%202021%20-%20WEF%20-%20UN%20SDGs %20Report_0.pdf.

Irwin, T. (2019). The emerging transition design approach. *Cuadernos del Centro de Estudios en Diseño y Comunicación. Ensayos*, 73, 147–179.

ITS. (2004). Caderno de debate: tecnologia social: direito à ciência e ciência para a cidadania. *Repositorio.mcti.gov.br.* https://repositorio.mcti.gov.br/han dle/mctic/5172.

Iwaniec, D. M., Cook, E. M., Barbosa, O., & Grimm, N. B. (2019). The framing of urban sustainability transformations. *Sustainability, 11*(3), 573.

Jennings, N., Fecht, D., & De Matteis, S. (2020). Mapping the co-benefits of climate change action to issues of public concern in the UK: a narrative review. *The Lancet Planetary Health, 4*(9), e424–e433.

Jia, S., Wang, Y., Wong, N. H., & Weng, Q. (2025). A hybrid framework for assessing outdoor thermal comfort in large-scale urban environments. Landscape and Urban Planning, 256, 105281. https://doi.org/10.1016/j .landurbplan.2024.105281

Judah, I. (2020). *Adaptive Mitigation: A Framework for Integrating Climate Adaptation and Mitigation Solutions in Urban Multi-unit Residential Buildings.* https://doi.org/10.14288/1.0395287.

Kaika, M. (2017). "Don't call me resilient again!": the New Urban Agenda as immunology … or … what happens when communities refuse to be vaccin- ated with 'smart cities' and indicators. Environment and Urbanization.

Kempeneer, S., Peeters, M., & Compernolle, T. (2021). Bringing the user Back in the building: an analysis of ESG in real estate and a behavioral framework to guide future research. *Sustainability, 13*(6), 3239.

Kern, K., & Alber, G. (2009). Governing Climate Change in Cities: Modes of Urban Climate Governance in Multi-Level Systems. In *The International Conference on Competitive Cities and Climate Change, Milan, Italy, 9 – 10 October, 2009* (pp. 171–196). OECD. https://edepot.wur.nl/51364

Kim, D., & Kang, J. E. (2018). Integrating climate change adaptation into community planning using a participatory process: the case of Saebat Maeul community in Busan, Korea. *Environment and Planning B: Urban Analytics and City Science*, *45*(4), 669–690.

Kindon, S., Pain, R., & Kesby, M. (2007). (Eds.). *Participatory Action Research Approaches and Methods*. London: Routledge.

Kirby, A. (2014). Adapting cities, adapting the curriculum. *Geography*, 99(2), 90–98.

Kirkby, P., Williams, C., & Huq, S. (2018). Community-based adaptation (CBA): Adding conceptual clarity to the approach, and establishing its principles and challenges. *Climate and Development*, *10*(7), 577–589.

Klimastadt Berlin. (2023). *Klimastadt Berlin 2030*. https://archplus.net/en/klimastadt-berlin-2030/

Klinsky, S., & Mavrogianni, A. (2020). Climate justice and the built environment. *Buildings and Cities*, *1*(1), 412–428.

Klinsky, S., & Sagar, A. D. (2022). The why, what and how of capacity building: some explorations. *Climate Policy*, *22*(5), 549–556.

Kohler, M. (2023). *Rain garden installations at Denton Place* [Photograph]. Personal collection.

Korpilo, S., Kaaronen, R. O., Olafsson, A. S., & Raymond, C. M. (2022). Public participation GIS can help assess multiple dimensions of environmental justice in urban green and blue space planning. *Applied Geography, 148,* 102794.

Leal Filho, W., Wall, T., Rui Mucova, S. A., Nagy, G. J., Balogun, A.-L., Luetz, J. M., Ng, A. W., Kovaleva, M., Safiul Azam, F. M., Alves, F., Guevara, Z., Matandirotya, N. R., Skouloudis, A., Tzachor, A., Malakar, K., & Gandhi, O. (2022). Deploying artificial intelligence for climate change adaptation. *Technological Forecasting and Social Change, 180,* 121662. https://doi.org/10.1016/j.techfore.2022.121662.

Leck, H. (2017). The role of culture in climate adaptation: 'the *nkanyamba* caused that storm.' *Third World Thematics: A TWQ Journal, 2*(2–3), 296–315. https://doi.org/10.1080/23802014.2017.1408424.

Lee, T., Yang, H., & Blok, A. (2020). Does mitigation shape adaptation? The urban climate mitigation-adaptation nexus. *Climate Policy, 20*(3), 341–353. https://doi.org/10.1080/14693062.2020.1730152.

Lefebvre, H. (1968). *Le droit à la ville*. Anthropos.

Lefebvre, H. (1996). The right to the city. *Writings on Cities*, 63181. https://freeebooksa7lr85.ml/0415534151-henri-lefebvre:-spatial-politics,-everyday-life-and-the-right-to-the-city.pdf.

Leonard, S., Parsons, M., Olawsky, K., & Kofod, F. (2013). The role of culture and traditional knowledge in climate change adaptation: insights from East Kimberley, Australia. *Global Environmental Change*, *23*(3), 623–632. https://doi.org/10.1016/j.gloenvcha.2013.02.012.

Leone, M., & Raven, J. (2018). Multi-scale and adaptive-mitigation design methods for climate resilient cities. *TECHNE-Journal of Technology for Architecture and Environment*, (15), 299–310. https://doi.org/10.13128/Techne-22076

Levidow, L., Sansolo, D., & Schiavinatto, M. (2021). Agroecological innovation constructing socionatural order for social transformation: two case studies in Brazil. *Tapuya: Latin American Science, Technology and Society*, *4*(1), 1843318. https://doi.org/10.1080/25729861.2020.1843318.

Levine, L. (2020). *NYC's New Plan Would Let Massive Sewage Overflows Continue*. NRDC. https://www.nrdc.org/bio/larry-levine/nycs-new-plan-would-let-massive-sewage-overflows-continue

Liao, K. H., Chan, J. K. H., & Huang, Y. L. (2019). Environmental justice and flood prevention: the moral cost of floodwater redistribution. *Landscape and Urban Planning*, *189*, 36–45.

Likhacheva Sokolowski, I., Maheshwari, A., Malik, A. (2019).

Green Buildings : A Finance and Policy Blueprint for Emerging Markets. Washington, D.C. : World Bank Group. http://documents.worldbank.org/curated/en/586841576523330833

Lindberg, F., Grimmond, C. S. B., Gabey, A., Huang, B., Kent, C. W., Sun, T., Theeuwes, N. E., Järvi, L., Ward, H. C., Capel-Timms, I., Chang, Y., Jonsson, P., Krave, N., Liu, D., Meyer, D., Olofson, K. F. G., Tan, J., Wästberg, D., Xue, L., & Zhang, Z. (2018). Urban Multi-scale Environmental Predictor (UMEP): An integrated tool for city-based climate services. *Environmental Modelling & Software*, *99*, 70–87. https://doi.org/10.1016/j.envsoft.2017.09.020.

Liu, K., Xue, M., Peng, M., & Wang, C. (2020). Impact of spatial structure of urban agglomeration on carbon emissions: an analysis of the Shandong Peninsula, China. *Technological Forecasting and Social Change*, *161*, 120313.

Long, J., & Rice, J. (2019). From sustainable urbanism to climate urbanism. *Urban Studies*, 56(5), 992–1008.

Lou, J., Hultman, N., Patwardhan, A., & Qiu, Y. L. (2022). Integrating sustainability into climate finance by quantifying the co-benefits and market impact of carbon projects. *Communications Earth & Environment, 3*(1), 137. https://doi.org/10.1038/s43247-022-00468-9.

Lucon O., D. Ürge-Vorsatz, A. Zain Ahmed, H. Akbari, P. Bertoldi, L. F. Cabeza, N. Eyre, A. Gadgil, L. D. D. Harvey, Y. Jiang, E. Liphoto, S. Mirasgedis, S. Murakami, J. Parikh, C. Pyke, & M. V. Vilariño. (2014). Buildings. In *Climate Change 2014: Mitigation of Climate Change. Contribution of Working Group III to the Fifth Assessment Report of the Intergovernmental Panel on Climate Change* [Edenhofer, O., R. Pichs-Madruga, Y. Sokona, E. Farahani, S. Kadner, K. Seyboth, A. Adler, I. Baum, S. Brunner, P. Eickemeier, B. Kriemann, J. Savolainen, S. Schlömer, C. von Stechow, T. Zwickel and J. C. Minx (eds.)]. Cambridge: Cambridge University Press.

Lwasa, S., Seto, K. C., Bai, X., Blanco, H., Gurney, K. R., Kilkiş, S., … & Yamagata, Y. (2022). Urban systems and other settlements. *Climate Change 2022: Mitigation of Climate Change. Contribution of Working Group III to the Sixth Assessment Report of the Intergovernmental Panel on Climate Change.*

Lynch, A. J. (2018). Creating effective urban greenways and stepping-stones: four critical gaps in habitat connectivity planning research. *Journal of Planning Literature, 34*(2), 131–155.

Magnan, A. (2014). Avoiding maladaptation to climate change: towards guiding principles. *SAPIENS, 7*(1), 1–10.

Mahendra, A., King, R., Du, J., Dasgupta, A., Beard, V., Kallergis, A., & Schalch, K. (2021). Towards a More Equal City: Seven Transformations for More Equitable and Sustainable Cities. *World Resources Institute.* https://doi.org/10.46830/wrirpt.19.00124.

Mahmoudi, D., Lubitow, A., & Christensen, M. A. (2019). Reproducing spatial inequality? The sustainability fix and barriers to urban mobility in Portland, Oregon. *Urban Geography, 41*(6), 801–822.

Marchezini, V., Wisner, B., Londe, L. R., & Saito, S. M. (2017). *Reduction of Vulnerability to Disasters: From Knowledge to Action*. São Paulo, Brazil:

RiMa Editora. Marco, C. (2024, March 5). *Are superblocks the future of urban living?* CNU. www.cnu.org/publicsquare/2024/03/05/are-superblocks-future-urban-living#:~:text=Finally%2C%20in%201993%2C%20the%20Urban.

Mayor of London. (2018). *Zero carbon London: A 1.5°C compatible plan.* www.london.gov.uk/sites/default/files/1.5_action_plan_amended.pdf.

McCauley, D., Ramasar, V., Heffron, R. J., Sovacool, B. K., Mebratu, D., & Mundaca, L. (2019). Energy justice in the transition to low carbon energy systems: exploring key themes in interdisciplinary research. *Applied Energy*, 233–234, 916–921.

McKibbin, W. J., & Wilcoxen, P. J. (2004). *Climate Policy and Uncertainty: The Roles of Adaptation versus Mitigation* (pp. 1–15). Washington, DC: Brookings Institution.

McPhearson, T., Pickett, S. T., Grimm, N. B., Niemelä, J., Alberti, M., Elmqvist, T., … & Qureshi, S. (2016). Advancing urban ecology toward a science of cities. *BioScience*, *66*(3), 198–212.

McPhearson, T., Iwaniec, D. M., Hamstead, Z. A., Berbés-Blázquez, M., Cook, E. M., Muñoz-Erickson, T. A., Mannetti, L., & Grimm, N. (2021). A Vision for Resilient Urban Futures. In Z. A. Hamstead, D. M. Iwaniec, T. McPhearson, M. Berbés-Blázquez, E. M. Cook, & T. A. Muñoz-Erickson (Eds.), *Resilient Urban Futures* (pp. 173–186). Springer International Publishing. https://doi.org/10.1007/978-3-030-63131-4_12.

McPhearson, T., Cook, E. M., Berbés-Blázquez, M., Cheng, C., Grimm, N. B., Andersson, E., Barbosa, O., Chandler, D. G., Chang, H., Chester, M. V., Childers, D. L., Elser, S. R., Frantzeskaki, N., Grabowski, Z., Groffman, P., Hale, R. L., Iwaniec, D. M., Kabisch, N., Kennedy, C., … Troxler, T. G. (2022). A social-ecological-technological systems framework for urban ecosystem services. *One Earth*, *5*(5), 505–518. https://doi.org/10.1016/j.oneear.2022.04.007.

McRae, I., Freedman, F., Rivera, A., Li, X., Dou, J., Cruz, I., Ren, C., Dronova, I., Fraker, H., & Bornstein, R. (2020). Integration of the WUDAPT, WRF, and ENVI-met models to simulate extreme daytime temperature mitigation strategies in San Jose, California. *Building and Environment*, *184*, 107180. https://doi.org/10.1016/j.buildenv.2020.107180.

Measham, T. G., Preston, B. L., Smith, T. F., Brooke, C., Gorddard, R., Withycombe, G., & Morrison, C. (2011). Adapting to climate change through local municipal planning: barriers and challenges. *Mitigation and Adaptation Strategies for Global Change*, *16*(8), 889–909. https://doi.org/10.1007/s11027-011-9301-2.

Milèn, A. (2001). *What Do We Know about Capacity Building? An Overview of Existing Knowledge and Good Practice*. Geneva: World Health Organisation.

Ministry of National Development. (2016). *Taking Actions Today for a Carbon-efficient Singapore*. www.mnd.gov.sg/docs/default-source/mnd-documents/publications-documents/climate-action-plan–take-action-today.pdf.

Mirzaei, P. A. (2021). CFD modeling of micro and urban climates: problems to be solved in the new decade. *Sustainable Cities and Society*, *69*, 102839. https://doi.org/10.1016/j.scs.2021.102839.

Mission Innovation (2023). *21st Century Climate Innovation Assessment. Identifying 1.5 °C compatible innovations in the 4th industrial revolution that can deliver what is needed to allow 11 billion people live flourishing lives.* www.misolutionframework.net/publications/21st-century-climate-innovation-assessment.

MIT. (2021). *Fast Forward: MIT's Climate Action Plan for the Decade.* https://climate.mit.edu/climateaction/fastforward.

Mohan, A.K., & Muraleedharan, G. (2025). Exploring co-production in redirecting climate urbanism. Climate and Development, 17(2), 134–152. https://doi.org/10.1080/17565529.2024.2339249

Mohtat, N., & Khirfan, L. (2021). The climate justice pillars vis-à-vis urban form adaptation to climate change: a review. *Urban Climate, 39*, 100951.

Montuori, B., Rosa, M., & Santos, M. C. (2017) Design by means of citizen activism: three cases illustrated by the action of Coletivo Maré, Rio de Janeiro, Brazil. *The Design Journal, 20*:sup1, S2973–S2990.

Moore, T., & Doyon, A. (2018). The uncommon nightingale: sustainable housing innovation in Australia. *Sustainability, 10*(10), 3469.

Mueller, N., Rojas-Rueda, D., Khreis, H., Cirach, M., Andrés, D., Ballester, J., Bartoll, X., Daher, C., Deluca, A., Echave, C., Milà, C., Márquez, S., Palou, J., Pérez, K., Tonne, C., Stevenson, M., Rueda, S., & Nieuwenhuijsen, M. (2020). Changing the urban design of cities for health: the superblock model. *Environment International, 134*, 105132. https://doi.org/10.1016/j.envint.2019.105132.

Murtagh, N., & Sergeeva, N. (2021). Agency and sustainability in the construction industry. In Teerikangas, S., Onkila, T., Koistinen, K., & Mäkelä, M. (Eds.). *Research handbook of sustainability agency.* Cheltenham, United Kingdom: Edward Elgar Publishing.

Nash, R. F. (1989). *The Rights of Nature: A History of Environmental Ethics.* Madison, WI: University of Wisconsin Press.

National Weather Service. (n.d.). *New York City, Central Park.* www.weather.gov/wrh/timeseries?site=knyc.

Nautiyal, S., & Klinsky, S. (2022). The knowledge politics of capacity building for climate change at the UNFCCC. *Climate Policy, 22*(5), 576–592.

Neary, M., Harrison, A., Crellin, G., Parekh, N., Saunders, G., Duggan, F., Williams, S., & Austin, S. (2010). *Learning Landscapes in Higher Education: Clearing Pathways, Making Spaces.* Involving academics in leadership, governance & the management of estates in higher education. University of Lincoln, DEGW, HEFCE, HEFW & Scottish Funding Council.

Neder, R. (2011). *Tecnologia social como pluralismo tecnológico 1.* Centro de Desenvolvimento Sustentável Observatório do Movimento pela Tecnologia

Social Universidade de Brasília, UnB. https://mediaserver.almg.gov.br/acervo/34/602/2034602.pdf.

New York Climate Exchange. (2021). *About US: The New York Climate Exchange*. https://nyclimateexchange.org/about-us/.

NOAA. (2025). *Daily Summaries Station Details*. https://www.ncdc.noaa.gov/cdo-web/datasets/GHCND/stations/GHCND:USW00094728/detail

Nocerino, G., & Leone, M. F. (2023). Computational LEED: computational thinking strategies and Visual Programming Languages to support environmental design and LEED credits achievement. *Energy and Buildings, 278*, 112626.

Nocerino, G., & Leone, M. F. (2024). WorkerBEE: A 3D Modelling Tool for Climate Resilient Urban Development. In *International Symposium: New Metropolitan Perspectives* (pp. 16-26). Springer, Cham.

Norman, B. (2018). Are autonomous cities our urban future? *Nature communications, 9*(1), 2111.

Norman, B., Newman, P., & Steffen, W. (2021). Apocalypse now: Australian bushfires and the future of urban settlements. *NPJ Urban Sustainability, 1*(1), 2.

Norman, D., & Stappers, P. J. (2016). DesignX: design and complex socio-technical systems. *She Ji: The Journal of Design, Economics, and Innovation, 1*(2), 83–106. https://doi.org/10.1016/j.sheji.2016.01.002

Noy, K., & Givoni, M. (2018). *Is "Smart Mobility" Sustainable? Examining the Views and Beliefs of Transport's Technological Entrepreneurs*. Sustainability: Science Practice and Policy, 10(2), 422.

NYC Mayor's Office of Climate and Environmental Justice. (2019). *OneNYC 2050 Building a Strong and Fair City-A Livable Climate*. https://climate.cityofnewyork.us/reports/onenyc-2050/.

NYC OpenData. (April 13, 2022). *DEP Infrastructure Program Map of NYC*. State of NJ, Esri, HERE, Garmin, USGS, NGA, EPA, NPS.

NYIT. (2019). *Urban Design Climate Workshop – From Climate Science to Climate Action: Gowanus, Brooklyn*. https://ulidigitalmarketing.blob.core.windows.net/ulidcnc/sites/54/2020/09/5f5bc22d59be9-5f5bc22d59beaULI-NY-Gowanus-UDCW-Report-Final-spreads.pdf.pdf.

NYIT SoAD Urban Design Climate Lab (2019), Climate-aligned scenario for Gowanus, Brooklyn, New York City [Drawing]. Graduate student group work from Urban Design Climate Lab, directed by Prof. Jeffrey Raven at New York Institute of Technology.

NYSERDA. (2019). NYStretch Energy Code – 2020. www.nyserda.ny.gov/-/media/Project/Nyserda/Files/Programs/Energy-Code-Training/NYStretch-Energy-Code-2020.pdf.

Oke, T. R., Mills, G., Christen, A., & Voogt, J. A. (2017). *Urban Climates*. Cambridge: Cambridge University Press.

Olazabal, M., Chu, E., Castán Broto, V., & Patterson, J. (2021). Subaltern forms of knowledge are required to boost local adaptation. *One Earth*, *4*(6), 828–838.

Oscilowicz, E., Anguelovski, I., García-Lamarca, M., Cole, H. V. S., Shokry, G., Perez-del-Pulgar, C., Argüelles, L., & Connolly, J. J. T. (2025). Grassroots mobilization for a just, green urban future: Building community infrastructure against green gentrification and displacement. Journal of Urban Affairs, 1–34. https://doi.org/10.1080/07352166.2023.2180381

Oscilowicz, E., Anguelovski, I., Triguero-Mas, M., García-Lamarca, M., Baró, F., & Cole, H. V. (2022). Green justice through policy and practice: a call for further research into tools that foster healthy green cities for all. *Cities & health*, *6*(5), 878–893.

Piggott-McKellar, A. E., Nunn, P. D., McNamara, K. E., & Sekinini, S. T. (2020). Dam(n) Seawalls: A Case of Climate Change Maladaptation in Fiji. In W. Leal Filho (Ed.), *Managing Climate Change Adaptation in the Pacific Region* (pp. 69–84). Cham, Switzerland: Springer International Publishing. https://doi.org/10.1007/978-3-030-40552-6_4.

Poljansek, K., Casajus Valles, A., Marin Ferrer, M., Artes Vivancos, T., Boca, R., Bonadonna, C., Branco, A., Campanharo, W., De Jager, A., De Rigo, D., Dottori, F., Durrant, T., Estreguil, C., Ferrari, D., Frischknecht, C., Galbusera, L., Garcia Puerta, B., Giannopoulos, G., Girgin, S … Wood, M. (2021). *Recommendations for National Risk Assessment for Disaster Risk Management in EU, EUR 30596 EN*. Luxembourg. 10.2760/80545.

Porter, L., Rickards, L., Verlie, B., Bosomworth, K., Moloney, S., Lay, B., Latham, B., Anguelovski, I., & Pellow, D. (2020). Climate Justice in a Climate Changed World. *Planning Theory & Practice*, *21*(2), 293–321. https://doi.org/10.1080/14649357.2020.1748959.

Pozzebon, M. (2015) Tecnologia social: a South American view of technology and society relationship. In F.-X. Vaujany, N. Mitev, G. F. Lanzara, A. Mukherjee (Eds.), *Materiality, Rules and Regulation, New Trends in Management and Organization Studies* (Chap. 1, 33–51). Palgrave Macmilian. https://doi.org/10.1057/9781137552648_2.

Preston-Jones, A. (2020). The importance of climate change education in urban planning: a review of planning courses at UK universities. In W. Leal Filho, G. Nagy, M. Borga, P. Chávez Muñoz & A. Magnuszewski (Eds.), *Climate Change, Hazards and Adaptation Options. Climate Change Management*. Cham: Springer

Prieur-Richard, A.-H., Walsh, B., Craig, M., Melamed, M. L., Pathak, M. Connors, S., Bai, X., Barau, A., Bulkeley, H., Cleugh, H., Cohen, M., Colenbrander, S., Dodman, D., Dhakal, S., Dawson, R., Greenwalt, J., Kurian, P., Lee, B., Leonardsen, L., Masson-Delmotte, V., Munshi, D., Okem, A., Delgado Ramos, G. C., Sanchez Rodriguez, R., Roberts, D., Rosenzweig, C., Schultz, S., Seto, K., Solecki, W., van Staden, M., & Ürge- Vorsatz, D. (2018). *Extended Version: Global Research and Action Agenda on Cities and Climate Change Science.* Edmonton, Canada: WCRP.

Puppim de Oliveira, J. A., Doll, C. N., & Suwa, A. (2013). *Urban development with climate co-benefits: aligning climate, environmental and other development goals in cities.* UNU-IAS Policy Report. United Nations University Institute of Advanced Studies (UNU-IAS), Yokohama.

Raven, J. (2019) From Climate Science to Practice, Urban Design & Climate Change, *Urban Design Journal*, Urban Design Group, London, ISSN: 1750 712X.

Raven, J. (2021). Urban Design Climate Workshops. In *World Scientific Encyclopedia of Climate Change: Case Studies of Climate Risk, Action, and Opportunity Volume 3.* 79–82.

Raven, J., Stone, B., Mills, G., Towers, J., Katzschner, L., Leone, M., … & Hariri, M. (2018). Urban planning and design. *Climate change and cities: Second assessment report of the urban climate change research network,* 139–172.

Raven, J., Braneon, C, and Esposito, M. (2020). *Urban Design Climate Workshop, from Climate Science to Climate Action: Gowanus, Brooklyn, 2019.* Urban Land Institute.

Raven, J. (2025), Framework for climate-aligned urban design [Drawing]. Author's personal collection.

Raven, J. (2019). Urban Climate Factors. Urban Design Group Journal, 49, pp. 22-24. ISSN: 1750 712X. [Drawing]. Author's personal collection.

Raven, J. (2025), Phasing: Urban Design Climate Workshop process [Drawing]. Author's personal collection.

Raymond, C. M., Frantzeskaki, N., Kabisch, N., Berry, P., Breil, M., Nita, M. R., Geneletti, D., and Calfapietra, C. (2017). A framework for assessing and implementing the co-benefits of nature-based solutions in urban areas. *Environmental Science and Policy,* 77:15–24.

Reckien, D., Lwasa, S., Satterthwaite, D., McEvoy, D., Creutzig, F., Montgomery, M., Schensul, D., Balk, D., and Khan, I. (2018a). Equity, environmental justice, and urban climate change. In Rosenzweig, C., W. Solecki, P. Romero-Lankao, S. Mehrotra, S. Dhakal, and S. AliIbrahim (Eds.), *Climate Change and Cities: Second Assessment Report of the Urban*

Climate Change Research Network. New York: Cambridge University Press, 173–224.

Reckien, D., Salvia, M., Heidrich, O., Church, J. M., Pietrapertosa, F., De Gregorio-Hurtado, S., … & Dawson, R. (2018b). How are cities planning to respond to climate change? Assessment of local climate plans from 885 cities in the EU-28. *Journal of Cleaner Production, 191*, 207–219.

Reid, H., & Huq, S. (2014). Mainstreaming community-based adaptation into national and local planning. *Climate and Development, 6*:4, 291–292.

Ren, C., Ng, E. Y.-Y., & Katzschner, L. (2011). Urban climatic map studies: a review. *International Journal of Climatology, 31*(15), 2213–2233.

RIBA. (2019). *Declaring a climate emergency – two years on the road to net zero.* www.ribaj.com/intelligence/riba-climate-emergency-two-years-on-alan-vallance.

Rice-Boayue, J., Garo, L., & Ilboudo Nebie, E. (2025). Community stakeholder perspectives for empowering EJ initiatives through Public Participation Geographic Information Systems (PPGIS). Environmental Research Letters, 20(2), 024057. https://doi.org/10.1088/1748-9326/ada6e0

Rigolon, A., Browning, M., Lee, K., & Shin, S. (2018). Access to urban green space in cities of the Global South: a systematic literature review. *Urban Science, 2*(3), 67.

Rigolon, A., Stewart, W. P., & Gobster, P. H. (2020). What predicts the demand and sale of vacant public properties? Urban greening and gentrification in Chicago. *Cities, 107*, 102948. https://doi.org/10.1016/j.cities.2020.102948.

Robin, E., & Broto, V. C. (2021). Towards a postcolonial perspective on climate urbanism. *International Journal of Urban and Regional Research, 45*(5), 869–878.

Rockström, J., W. Steffen, K. Noone, Å. Persson, F. S. Chapin, III, E. Lambin, T. M. Lenton, M. Scheffer, C. Folke, H. Schellnhuber, B. Nykvist, C. A. De Wit, T. Hughes, S van der Leeuw, H. Rodhe, S. Sörlin, P. K. Snyder, R. Costanza, U. Svedin, M. Falkenmark, L. Karlberg, R. W. Corell, V. J. Fabry, J. Hansen, B. Walker, D. Liverman, K. Richardson, P. Crutzen, & J. Foley, (2009). Planetary boundaries: exploring the safe operating space for humanity. *Ecol. Soc., 14*, no. 2, 32.

Roggema, R., L. Vos, & J. Martin (2017) Community based environmental design: empowering local expertise in design charrettes. In W. Yan & W. Galloway (Eds.) *Rethinking Resilience. Innovation and Adaptation in a Time of Change.* Cham: Switzerland: Springer, 321–340.

Roggema, R. (Ed.). (2023). *Trends in Urban Design: Insights for the Future Urban Professional.* Cham, Switzerland: Springer International Publishing. https://doi.org/10.1007/978-3-031-21456-1.

Romero-Lankao, P., Bulkeley, H., Pelling, M., Burch, S., Gordon, D. J., Gupta, J., Johnson, C., Kurian, P., Lecavalier, E., Simon, D., Tozer, L., Ziervogel, G., & Munshi, D. (2018). Urban transformative potential in a changing climate. *Nature Climate Change, 8*(9), 754–756. https://doi.org/10.1038/s41558-018-0264-0.

Rosenthal, J. K., Sclar, E. D., Kinney, P. L., Knowlton, K., Crauderueff, R., & Brandt-Rauf, P. W. (2007). Links between the built environment, climate and population health: interdisciplinary environmental change research in New York City. *Annals-Academy of Medicine Singapore, 36*(10), 834.

Rosenzweig, C., Parry, M., & De Mel, M. (2021). *Our Warming Planet: Climate Change Impacts and Adaptation* (Vol. 02). World Scientific. https://doi.org/10.1142/12312.

Rosenzweig, C., Solecki, W., Hammer, S. A., & Mehrotra, S. (Eds.). (2011). *Climate Change and Cities: First Assessment Report of the Urban Climate Change Research Network*. Cambridge: Cambridge University Press.

Rosenzweig, C., Solecki, W., Pathak, M., Barau, A. S., Barata, M., & Dombrov, M. (2025). Clearing roadblocks that stymie city climate action. Nature Cities. https://doi.org/10.1038/s44284-025-00205-1

Rosenzweig, C., Solecki, W. D., Romero-Lankao, P., Mehrotra, S., Dhakal, S., & Ali Ibrahim, S. (2018). *Climate Change and Cities: Second Assessment Report of the Urban Climate Change Research Network*. In C. Rosenzweig, W. D. Solecki, P. Rotter, M., Hoffmann, E., Hirschfeld, J., Schröder, A., Mohaupt, F., and Schäfer, L. (2018). *Stakeholder participation in adaptation to climate change-lessons and experience from Germany*. Technical report, Institute for Ecological Economy Research.

Ross, T. (2013). The Commons, Melbourne. www.nightingalehousing.org/project/the-commons.

Rotter, M., Hoffman, E., Hirschfeld, J., Schröder, A., Mohaupt, F., & Schäfer, L. (2013). *Stakeholder Participation in Adaptation to Climate Change – Lessons and Experience from Germany*. OSTI. www.osti.gov/etdeweb/servlets/purl/22138775.

Takaes Santos, I. (2020). Confronting governance challenges of the resource nexus through reflexivity: a cross-case comparison of biofuels policies in Germany and Brazil. *Energy Research & Social Science, 65*, 101464. https://doi.org/10.1016/j.erss.2020.101464.

Santos, A. S., de Abreu, V. H. S., de Assis, T. F., Ribeiro, S. K., & Ribeiro, G. M. (2021). An overview on costs of shifting to sustainable road transport: a challenge for cities worldwide. In: Muthu, S.S. (eds) Carbon Footprint Case Studies. Environmental Footprints and Eco-design of Products and

Processes. Springer, Singapore, 93–121. https://doi.org/10.1007/978-981-15-9577-6_4.

Satorras, M., Ruiz-Mallén, I., Monterde, A., & March, H. (2020). Co-production of urban climate planning: insights from the Barcelona Climate Plan. *Cities*, *106*, 102887.

Schipper, E. L., Ayers, J., Reid, H., Huq, S., & Rahman, A. (2014). *Community-Based Adaptation to Climate Change: Scaling It Up*. Routledge.

Schleussner, C.-F., Ganti, G., Lejeune, Q., Zhu, B., Pfleiderer, P., Prütz, R., Ciais, P., Frölicher, T. L., Fuss, S., Gasser, T., Gidden, M. J., Kropf, C. M., Lacroix, F., Lamboll, R., Martyr, R., Maussion, F., McCaughey, J. W., Meinshausen, M., Mengel, M., … Rogelj, J. (2024). Overconfidence in climate overshoot. *Nature, 634*(8033), 366–373. https://doi.org/10.1038/s41586-024-08020-9

Schot, J., Brand, E., & Fischer, K. (1997) The greening of industry for a sustainable future: building an international research agenda. *Business Strategy and the Environment*, 6, 3, 153–162.

Schüle, S. A., Hilz, L. K., Dreger, S., & Bolte, G. (2019). Social inequalities in environmental resources of green and blue spaces: a review of evidence in the WHO European region. *International Journal of Environmental Research and Public Health*, *16*(7), 1216.

Scorza, F., & Santopietro, L. (2024). A systemic perspective for the Sustainable Energy and Climate Action Plan (SECAP). European Planning Studies, 32(2), 281–301. https://doi.org/10.1080/09654313.2021.1954603.

Shammin, M. R., Haque, A. E., & Faisal, I. M. (2022). A framework for climate resilient community-based adaptation. *Climate Change and Community Resilience*, 11–30. Springer Nature, Singapore. https://doi.org/10.1007/978-981-16-0680-9_2.

Sharifi, A. (2021). Co-benefits and synergies between urban climate change mitigation and adaptation measures: A literature review. Science of the total environment, 750, 141642.

Sharifi, A., & Yamagata, Y. (2015). A conceptual framework for assessment of urban energy resilience. *Energy Procedia*, *75*, 2904–2909.

Shi, L. (2021). From progressive cities to resilient cities: Lessons from history for new debates in equitable adaptation to climate change. *Urban Affairs Review*, *57*(5), 1442–1479.

Shi, L., Chu, E., Anguelovski, I., Aylett, A., Debats, J., Goh, K., … Van Deveer, S. D. (2016). Roadmap towards justice in urban climate adaptation research. *Nature Climate Change*, *6*(2), 131.

Shi, Y., Bilal, M., Ho, H. C., & Omar, A. (2020). Urbanization and regional air pollution across South Asian developing countries: a nationwide land use regression for ambient PM2.5 assessment in Pakistan. *Environ Pollut*, 266(Pt 2), 115145.

Shrestha, S. and Shobhakar D. (2019). An assessment of potential synergies and trade-offs between climate mitigation and adaptation policies of Nepal. *Journal of Environmental Management*, 235: 535–545.

Singh, P. K., & Chudasama, H. (2021). Pathways for climate resilient development: Human well-being within a safe and just space in the 21st century. *Global Environmental Change, 68*, 102277.

Solecki, W., Seto, K. C., Balk, D., Bigio, A., Boone, C. G., Creutzig, F., . . . & Zwickel, T. (2015). A conceptual framework for an urban areas typology to integrate climate change mitigation and adaptation. *Urban Climate, 14*, 116–137.

Solecki, W., Pathak, M., Barata, M., Salisu Barau, A., Dombrov, M., & Rosenzweig, C. (Eds.). (in press). *Climate Change and Cities: Third Assessment Report of the Urban Climate Change Research Network.* Cambridge: Cambridge University Press.

Sovacool, B. K., Kester, J., Noel, L., & de Rubens, G. Z. (2019). Energy injustice and Nordic electric mobility: inequality, elitism, and externalities in the electrification of vehicle-to-grid (V2G) transport. *Ecological economics, 157*, 205–217.

Sovacool, B. K., Barnacle, M. L., Smith, A., & Brisbois, M. C. (2022). Towards improved solar energy justice: Exploring the complex inequities of household adoption of photovoltaic panels. *Energy Policy, 164*, 112868.

Spires, M., Shackleton, S., Cundill, G. (2014) Barriers to implementing planned community based adaptation in developing countries: a systematic literature review. *Climate and Development* 6(3), 277–287.

Suckall, N., & Tompkins, E. L. (2020). Climate compatible development: generating co-benefits from climate change planning. *Sustainability, 12*(2), 496. https://doi.org/10.3390/su12020496.

Sun, F., Chen, P., & Jiao, J. (2018). Promoting public bike-sharing: a lesson from the unsuccessful Pronto system. *Transportation Research Part D: Transport and Environment, 63*, 533–547.

TAHMO. (2025). *Trans-African Hydro-Meteorological Observatory.* https://tahmo.org/

Tan, L. M., Arbabi, H., Densley Tingley, D., Brockway, P. E., & Mayfield, M. (2021). Mapping resource effectiveness across urban systems. *NPJ Urban Sustainability, 1*(1), 20. https://doi.org/10.1038/s42949-020-00009-3.

Teicher, H. M. (2018). Practices and pitfalls of competitive resilience: urban adaptation as real estate firms turn climate risk to competitive advantage. *Urban Climate, 25*, 9–21.

Temple, J. (2024, July 17). *Google, Amazon and the problem with Big Tech's climate claims*. MIT Technology Review. www.technologyreview.com/2024/07/17/1095019/google-amazon-and-the-problem-with-big-techs-climate-claims/.

Tersigni, E. (2019). *Community engagement session: outlining synergies between community and climate priorities for the Durban Isipingo central business district* [Photograph]. Personal collection.

Thome, A. M. T., Ceryno, P. S., Scavarda, A., & Remmen, A. (2016). Sustainable infrastructure: a review and a research agenda. *Journal of Environmental Management*, 184, 2, 143–156.

Thompson, S. M., & Capon, A. G. (2010). Designing a healthy and sustainable future: a vision for interdisciplinary education, research and leadership. In *Proceedings of the 2nd International Conference on Design Education (ConnectEd'10)*.

Tol, R. S. (2005). The marginal damage costs of carbon dioxide emissions: an assessment of the uncertainties. *Energy Policy*, *33*(16), 2064–2074.

Toparlar, Y., Blocken, B., Maiheu, B., & Van Heijst, G. J. F. (2017). A review on the CFD analysis of urban microclimate. *Renewable and Sustainable Energy Reviews*, *80*, 1613–1640. https://doi.org/10.1016/j.rser.2017.05.248.

Torabi, E., Dedekorkut-Howes, A., & Howes, M. (2018). Adapting or maladapting: building resilience to climate-related disasters in coastal cities. *Cities*, 72, 295–309.

Towers, J. (2005, July/August). *Sustainable design*. www.euro-peanoutdoorgroup.com/downloads/sustainable-design/Sustainabledesign.pdf.

Tozer, L., & Klenk, N. (2018). Discourses of carbon neutrality and imaginaries of urban futures. *Energy Research & Social Science*, *35*, 174–181.

Tribbia, J., & Moser, S. (2008). More than information: what coastal managers need to plan for climate change *Environmental Science & Policy 11*(4), 315–328.

UK House of Commons. (2008). *Planning Matters – Labour Shortages and Skills Gaps: 11th Report of Session 2007–08*, Vol. 2: Oral and Written Evidence, Volume 2, HC517–2.

UK House of Commons Environmental Audit Committee. (2021). *Green Jobs*. Third Report of Session 2021–22, HC75.

UK Met Office. (2011). *Weather Observations Website (WOW)*. www.metoffice.gov.uk/weather/learn-about/met-office-for-schools/other-content/other-resources/are-you-obsessed-with-the-weather#:~:text=The%20Weather%20Observations%20Website%20(WOW,or%20the%20frequency%20of%20reports.

UK Parliamentary Committee on Climate Change, (2023). *Progress in reducing emissions*. 2023 Report to Parliament.

UN. (1992). *United Nations Conference on Environment & Development Rio de Janerio, Brazil, 3 to 14 June 1992*. https://sustainabledevelopment.un.org/content/documents/Agenda21.pdf.

United Nations. (2023a). Capacity-building at the systemic, institutional and individual levels. www.unfccc.int.

United Nations. (2023b). Paris Committee on Capacity-building (PCCB). www.unfccc.int.

University of Minnesota Twin Cities. (2018). *Resilient Adaptation of Sustainable Buildings*. Center for Sustainable Building Research. Prepared For the Minnesota Pollution Control Agency: Laura Millberg, Sustainable Development and Climate Resilience Planner MPCA. https://design.umn.edu/sites/design.umn.edu/files/2020-10/csbr_mpca_resilience_report_final_2018.pdf

UNEP and UN-Habitat. (2021). *GEO for Cities – Towards Green and Just Cities*. www.unep.org/resources/report/geo-cities-towards-green-and-just-cities.

UNEP. (2022). *Adaptation Gap Report 2022: Too Little, Too Slow Climate: Adaptation Failure Puts World at Risk*. UN.

UNEP. (2023). Adaptation Gap Report 2023: Underfinanced. Underprepared. Inadequate investment and planning on climate adaptation leaves world exposed. Nairobi. https://doi.org/10.59117/20.500.11822/43796.

UNEP, UN-Habitat, & World Bank. (2010). *International standard for determining greenhouse gas emissions from cities*. Available at: www.siteresources.worldbank.org.

UNDP. (1998). *Capacity Assessment and Development – in a systems and strategic management context*.

UNDP. (2022). *New Threats to Human Security in the Anthropocene: Demanding Greater Solidarity*. United Nations Development Programme. https://bit.ly/45GEzdJ.

UNESCO. (2017). *UN World Water Development Report 2017 – Wastewater, the Untapped Resource*. www.unesco.org/en/wwap/wwdr/2017.

UNFCCC. (2015). *Paris Agreement*. https://unfccc.int/files/meetings/paris_nov_2015/application/pdf/paris_agreement_english_.pdf.

UNFCCC. (2020). *Annual Report*. https://unfccc.int/about-us/annual-report/annual-report-2020.

UNFCCC. (2021). *Glasgow Pact*. https://unfccc.int/sites/default/files/resource/cma2021_10_add1_adv.pdf.

UNFCCC. (2022). *China's Achievements, New Goals and New Measures for Nationally Determined Contributions.* https://unfccc.int/sites/default/files/NDC/2022-06/China%E2%80%99s%20Achievements%2C%20New%20Goals%20and%20New%20Measures%20for%20Nationally%20Determined%20.

UNFCCC (2023) *Methodologies for assessing adaptation needs and their application Summary.* https://unfccc.int/sites/default/files/resource/ac22_6a_assess_needs.pdf.

UN-DESA. (2020). *Annual Highlights Report.* www.un.org/en/desa/highlights-report-2020-2021.

UN-Habitat (Ed.). (2020). *The new urban agenda.* https://unhabitat.org/sites/default/files/2020/12/nua_handbook_14dec2020_2.pdf.

UN-Habitat. (2020). *World Cities Report 2020: The value of sustainable urbanization.* United Nations Human Settlements Programme.

UNEP-PRI. (2021). *Principles for responsible investment. Annual report 2021.* UNEP. www.unpri.org.

Urban Sustainability Directors Network. (2017). *Guide to equitable, community-driven climate preparedness planning.*

USDA. (2024). *New York City Urban Field Station.* https://research.fs.usda.gov/nrs/centers/nyc.

USGBC. (2025). *LEED v5.* https://www.usgbc.org/leed/v5

Venter, Z. S., Figari, H., Krange, O., & Gundersen, V. (2023). Environmental justice in a very green city: spatial inequality in exposure to urban nature, air pollution and heat in Oslo, Norway. *Science of The Total Environment, 858,* 160193.

Ville de Paris. (2018). *Paris Climate Action Plan-Towards a Carbon Neutral City and 100& Renewable Energy.* https://cdn.paris.fr/paris/2020/11/23/a10afc931be2124e21e39a1624132724.pdf.

Visseren-Hamakers, I. J., Razzaque, J., McElwee, P., Turnhout, E., Kelemen, E., Rusch, G. M., . . . Zaleski, D. (2021). Transformative governance of biodiversity: insights for sustainable development. *Current Opinion in Environmental Sustainability,* 53, 20–28.

Wachsmuth, D., Basalaev-Binder, R., Pcae, N., & Seltz, L. (2019). Bridging the boroughs: how well does New York's bike sharing system serve New Yorkers. *Urban Planning and Governance Research Group, McGill University, School of Urban Planning, Montreal, Canada.*

Wai, C. Y., Chau, H.-W., Paresi, P., & Muttil, N. (2025). Experimental Analysis of Cool Roof Coatings as an Urban Heat Mitigation Strategy to Enhance Thermal Performance. Buildings, 15(5), 685. https://doi.org/10.3390/buildings15050685

Walsh, C. L., Dawson, R. J., Hall, J. W., Barr, S. L., Batty, M., Bristow, A. L., Carney, S., Dagoumas, A. S., Ford, A. C., Harpham, C., Tight, M. R., Watters, H., & Zanni, A. M. (2011). Assessment of climate change mitigation and adaptation in cities. *Proceedings of the Institution of Civil Engineers – Urban Design and Planning, 164*(2), 75–84. https://doi.org/10.1680/udap.2011.164.2.75.

Wang, S. H., Huang, S. L., & Huang, P. J. (2018). Can spatial planning really mitigate carbon dioxide emissions in urban areas? A case study in Taipei, Taiwan. *Landscape and Urban Planning, 169*, 22–36.

Warren-Myers, G., Hürlimann, A., & Bush, J. (2020). Advancing capacity to adapt to climate change in the Australian property industry – addressing climate change information needs. *Journal of European Real Estate Research, 13*(3), 321–335. https://doi.org/10.1108/JERER-03-2020-0017

Watkiss, P., Benzie, M., & Klein, R. J. T. (2015). The complementarity and comparability of climate change adaptation and mitigation. *Wiley Interdisciplinary Reviews: Climate Change, 6*(6), 541–557.

Webb, R., Bai, X., Smith, M. S., Costanza, R., Griggs, D., Moglia, M., . . . & Thomson, G. (2018). Sustainable urban systems: co-design and framing for transformation. *Ambio, 47*, 57–77.

Wiedmann, T., Chen, G., Owen, A., Lenzen, M., Doust, M., Barrett, J., & Steele, K. (2021). Three-scope carbon emission inventories of global cities. *Journal of Industrial Ecology,* 25(3), 735–750.

Wolfram, M., Borgström, S. & Farrelly, M. (2019). Urban transformative capacity: from concept to practice. *Ambio* 48, 437–448 (2019).

World Bank. (2022). *Annual Report 2022. Helping Countries Adapt to a Changing World*. The World Bank.

World Resources Institute (WRI) and the World Business Council for Sustainable Development (WBCSD) (2004). Greenhouse Gas Protocol: a Corporate Accounting and Reporting Standard. Revised edition. WRI, Geneva, Switzerland and WBCSD, Washington DC, USA.

Wyborn, C., Datta, A., Montana, J., Ryan, M., Leith, P., Chaffin, B., Miller, C., & van Kerkhoff, L. (2019). Co-Producing Sustainability: reordering the governance of science, policy, and practice. *Annual Review of Environment and Resources, 44*(1), 319–346.

Yagoub, M. M., Tesfaldet, Y. T., Elmubarak, M. G., & Al Hosani, N. (2022). Extraction of urban quality of life indicators using remote sensing and machine learning: the case of Al Ain City, United Arab Emirates (UAE). *ISPRS International Journal of Geo-Information, 11*(9), 458.

Yenneti, K., & Day, R. (2016). Distributional justice in solar energy development in India: the case of Charanka solar park. *Journal of Rural Studies*, *46*, 35–46. https://doi.org/10.1016/j.jrurstud.2016.05.009.

Yigitcanlar, T., Mehmood, R., & Corchado, J. M. (2021). Green artificial intelligence: towards an efficient, sustainable and equitable technology for smart cities and futures. *Sustainability*, 13, 16, 8952.

Yin, S., Ren, C., Zhang, X., Hidalgo, J., Schoetter, R., Ting Kwok, Y., & Ka-Lun Lau, K. (2022)."Potential of Synthetizing Climatopes and Local Climate Zones for Urban Climatic Planning Recommendations: A Case Study in Toulouse, France." Cybergeo: European Journal of Geography. http://journals.openedition.org/cybergeo/39417.

Yuan, C., Zhu, R., Tong, S., Mei, S., & Zhu, W. (2022). Impact of anthropogenic heat from air-conditioning on air temperature of naturally ventilated apartments at high-density tropical cities. *Energy and Buildings*, *268*, 112171. https://doi.org/10.1016/j.enbuild.2022.112171.

Yuan, X., Liang, Y., Hu, X., Xu, Y., Chen, Y., & Kosonen, R. (2023). Waste heat recoveries in data centers: a review. *Renewable and Sustainable Energy Reviews*, *188*, 113777. https://doi.org/10.1016/j.rser.2023.113777.

Zare, S., Hasheminejad, N., Shirvan, H. E., Hemmatjo, R., Sarebanzadeh, K., & Ahmadi, S. (2018). Comparing Universal Thermal Climate Index (UTCI) with selected thermal indices/environmental parameters during 12 months of the year. *Weather and climate extremes*, 19, 49–57.

Zhao, D., Cai, J., Xu, Y., Liu, Y., & Yao, M. (2023). Carbon sinks in urban public green spaces under carbon neutrality: a bibliometric analysis and systematic literature review. *Urban Forestry & Urban Greening*, 128037.

Ziervogel, G., Cowen, A., & Ziniades, J. (2016). Moving from adaptive to transformative capacity: building foundations for inclusive, thriving, and regenerative urban settlements. *Sustainability: Science Practice and Policy*, 8(9), 955.

Zigurat Institute of Technology (2021). *Map of Barcelona Superblocks*. www.e-zigurat.com/en/blog/tactical-urban-planning-superblocks-eixample-district-barcelona/

Znidarcic, T. (2015). *Kano, Nigeria* [Photograph]. Tadej Znidarcic Photography. www.tadejznidarcic.com/.

Zukin, S. (2020). *The innovation complex: Cities, tech, and the new economy.* Oxford University Press.

Acknowledgments

This material is based upon work supported by the National Science Foundation City-as-Lab INFEWS/T3 RCN Supplement under Grant No. 1856032 – Amendment ID 003.

This publication has received support by the Cooperation Partnership for Higher Education (KA220 HED) "UCCRN_edu – Climate Resilient Design, Planning and Governance of Cities" funded by the Erasmus Plus program of the European Union under the Grant Agreement 2021–1-IT02-KA220-HED-000027520.

We thank UCCRN Hub Directors – Ken Doust (Australia Oceania Hub), Martin Lehmann (Nordic Node), Franco Montalto (North America Hub), and Juan Camilo Osorio (Latin America Hub) – for their excellent shepherding of this Element. Jaad Benhallam and Natalie Kozlowski, of the UCCRN Secretariat provided outstanding editorial and graphic support. We greatly appreciate the constructive reviews of the Element by John Black, Catherine Brinkley, Andrea Santos, and Amanda Yates.

Author List

Coordinating Lead Authors

Jeffrey Raven, *New York Institute of Technology, New York*
Mattia Federico Leone, *Università di Napoli Federico II, Naples*

Lead Authors

Sanjukkta Bhaduri, *School of Planning and Architecture, Delhi*
Christian Braneon, *Columbia Climate School / NASA Goddard Institute for Space Studies, New York*
David Corbett, *ICLEI, Munich*
David Driskell, *Community Planning Collaborative, Seattle*
Ursula Eicker, *Concordia University, Montréal*
John Fernández, *MIT, Cambridge*
Jing Gan, *Tongji University and Ministry of Natural Resources of China, Shanghai*
Anna Hürlimann, *The University of Melbourne, Melbourne*
Ilana Judah, *ARUP, New York*
Michael Neuman, *University of Westminster, London*
Barbara Norman, *University of Canberra and Australian National University, Canberra*
Dennis Pamlin, *Mission Innovation NCI / RISE, Stockholm*
Chao Ren, *The University of Hong Kong, Hong Kong*
Rob Roggema, *Monterrey Institute of Technology and Higher Education, Monterrey*
Pourya Salehi, *ICLEI, Bonn*
Anne Shellum, *Google Sustainability, Denver*
Andréa Souza Santos, *COPPE – Federal University of Rio de Janeiro*
Joel Towers, *Parsons School of Design / The New School, New York*
Cristina Visconti, *Università di Napoli Federico II, Naples*

Contributing/Case Study Authors

Sara Artaega-Morales, *Universidad de Antioquia, Medellín*
Ione Avila-Palencia, *Queen's University Belfast, Belfast*
Jole Lutzu, *ICLEI Europe, Barcelona*
Bruno Barroca, *Gustave Eiffel University, Paris*
Martina Kohler, *Parsons School of Design, New York*

Margot Pellegrino, *Gustave Eiffel University, Paris*

Juliana Velez Duque, *C40 Cities, Medellín*

Darien Alexander Williams, *Boston University Institute for Global Sustainability (IGS), Boston*

Sean O'Donoghue, *eThekwini Municipality, Durban*

Maria Iaccarino, *Naples Municipality, Naples*

Enza Tersigni, *Università di Napoli Federico II, Naples*

Aliyu Salisu Barau, *Bayero University, Kano*

Enjoli Dominique Hall, *Massachusetts Institute of Technology (MIT), Cambridge*

Daphne Gondhalekar, *Technical University of Munich, Munich*

Element Shepherds

Ken Doust, *Southern Cross University, Lismore*

Martin Lehmann, *Aalborg University, Aalborg*

Franco Montalto, *Drexel University, Philadelphia*

Juan Camilo Osorio, *Pratt Institute, New York*

Element Scientists

Enza Tersigni, *Università di Napoli Federico II, Naples*

Mia Prall, *Aalborg University, Aalborg*

Priska Marianne, *Independent Consultant, Jakarta*

Colin Marchenoir, *SOCOTEC UK, London*

Climate Change and Cities: Third Assessment Report of the Urban Climate Change Research Network

Series Editors

William Solecki
New York

William Solecki is a Professor in the Department of Geography at Hunter College, City University of New York (CUNY). From 2006–2014 he served as the Director of the CUNY Institute for Sustainable Cities at Hunter College. He also served as interim Director of the Science and Resilience Institute at Jamaica Bay. He has co-led several climate impacts studies in the greater New York and New Jersey region, including the New York City Panel on Climate Change (NPCC). He was a Lead Author of the Urban Areas chapter in the Fifth Assessment Report of the Intergovernmental Panel on Climate Change (IPCC), and a Coordinating Lead Author of the Urbanization, Infrastructure, and Vulnerability chapter in the Third National Climate Assessment (US). He is a co-founder of the Urban Climate Change Research Network (UCCRN), co-editor of Current Opinion on Environmental Sustainability, and founding editor of the Journal of Extreme Events. His research focuses on urban environmental change, resilience, and adaptation transitions.

Minal Pathak
Ahmedabad

Minal Pathak is an Associate Professor at the Global Centre for Environment and Energy at Ahmedabad University, India. She is a Senior Scientist with the Technical Support Unit of Working Group III of the IPCC for its Sixth Assessment cycle. She has contributed to two IPCC Special Reports, co-edited the IPCC Sixth Assessment Report, and contributed to the recently published IPCC Sixth Assessment Synthesis Report. She heads the South Asia Hub of the UCCRN, headquartered at the Columbia Climate School. She was a Visiting Researcher at Imperial College London (2017–2023) and a Visiting Scholar at MIT (2016–2017). Her research focuses on climate change mitigation strategies for urban settlements, transport, and buildings, and their co-benefits/interlinkages with development.

Martha Barata
Rio de Janeiro

Martha Barata is Coordinator of the Latin America Hub of the UCCRN, headquartered at Columbia Climate School. Barata is a collaborating researcher at the Oswaldo Cruz Institute (Fiocruz) and CentroClima (COPPE/UFRJ), following retirement from the Oswaldo Cruz Foundation in 2017. She was a Visiting Scholar in the Center for Climate Systems Research at Columbia University in 2014, conducting research on climate risk management in cities.

Aliyu Salisu Barau
Kano

Aliyu Salisu Barau is a Professor in Urban Development and Management at the Department of Urban and Regional Planning and Fifth Dean of the Faculty of Earth and Environmental Sciences at Bayero University in Kano, Nigeria. He is a transdisciplinary researcher with interests in climate change, landscape ecology, clean energy, socio-ecological systems, sustainability agenda setting, informally and formally protected ecosystems, special economic zones, and inclusive and innovative planning. He contributes to the research, policy, and action agenda in Nigeria and globally through engagements with UN Environment, IPCC, Future Earth, IUCN, IPBES, IIED, UNICEF, and UN Habitat. He is also the director of the West Africa Center for the UCCRN at Columbia University in New York.

Maria Dombrov

New York

Maria Dombrov is a Senior Research Associate I at the Climate Impacts Group, co-located at NASA Goddard Institute for Space Studies and Columbia University's Center for Climate Systems Research, in New York City. Ms. Dombrov is UCCRN's Global Coordinator and the Project Manager of UCCRN's Third Assessment Report on Climate Change and Cities (ARC3.3). Her work focuses on understanding the risks and vulnerabilities that climate change and extreme events present to cities and their metropolitan regions around the world.

Cynthia Rosenzweig

New York

Cynthia Rosenzweig is a Senior Research Scientist at the NASA Goddard Institute for Space Studies (GISS), Adjunct Senior Research Scientist at the Columbia University Climate School, and Adjunct Professor in the Department of Environmental Science at Barnard College. At NASA GISS, she heads the Climate Impacts Group, whose mission is to investigate the interactions of climate on systems and sectors important to human well-being.

Dr. Rosenzweig is Co-Founder and Co-Director of the Urban Climate Change Research Network (UCCRN). She was Co-Chair of the New York City Panel on Climate Change (NPCC). Dr. Rosenzweig has served as Coordinating Lead Author and Lead Author on several IPCC Assessment Reports.

About the Series

This Elements series, published in collaboration with the Urban Climate Change Research Network (UCCRN), provides essential knowledge on climate change and cities for researchers, practitioners, policymakers, and students. Bridging the gap between theory and practice, the series invites readers to engage with the latest advances in the field. Focusing on urban transformation, cross-cutting themes, and urgent research areas, it empowers stakeholders to drive impactful climate action in rapidly-evolving urban contexts.

Printed by Integrated Books International,
United States of America